The specialist DECORATOR

How to gain profit and success in decorating by specialising

Pete Wilkinson

First published in the United Kingdom in 2020

Copyright © Peter Wilkinson

Peter Wilkinson has asserted his right to be identified as the author of this work in accordance with the Copyright, Designs and Patents Act 1988

All rights reserved. No part of this publication may be reproduced, stored in a retrieval system, or transmitted in any form or by any means, electronic, mechanical, photocopying, recording or otherwise, without prior permission of the copyright owner.

This book is not intended to provide personalised legal, financial, or investment advice. The Author and Publisher specifically disclaim any liability, loss or risk which is incurred as a consequence, directly or indirectly of the use and application of any contents of this work.

2020 edition

First edition

26-11-2020

This book is dedicated to all those working-class dreamers who by sheer graft get there in the end.

We are born to be free and freedom should be our ultimate goal.

Contents

Section 1 – Business pointers when specialising

Chapter 1 – Introduction

Why is a decorating specialism the way to go?

Chapter 2 – Breaking away from the herd

Carve out your own little bit of the industry and then dominate it.

Chapter 3 – Your name, tagline and pitch

Making those important decisions at the very start of your journey.

Chapter 4 – Barriers to entry

These are more important than you think.

Contents

Section 2 – The basics

Chapter 5 – Domestic interiors

We start off simple with a look at decorating Mrs Jones's lounge.

Chapter 6 – Domestic exteriors

You have to dodge the weather, but good money can be made from this specialism that a lot of decorators avoid.

Chapter 7 – Paperhanging

Rapidly coming back into fashion and with few decorators brave enough to do it, this can be a great specialism.

Chapter 8 – Fences

It sounds like a simple idea, but this could be a profitable and scalable specialism.

Chapter 9 – Churches

A little off the radar and they don't get painted often but when they do few companies can handle the work.

Contents

Section 3 – Spraying

Chapter 10 – Spraying overview

My specialism, so I had to dedicate a section to it.

Chapter 11 – Apartments

This specialism is big and if you get into this work it can pay well and give you a steady stream of work.

Chapter 12 – Shutters

A profitable and saleable business if you spray.

Chapter 13 – Boats

Another one that is off the radar for most decorators but can pay well.

Chapter 14 – Spray plaster

A new specialism in this country for decorators.

Contents

Section 4 – The moneymakers

Chapter 15 – Computer-aided signmaking

We used to do signwriting in the past, this is the modern equivalent.

Chapter 16 – Intumescent coatings

This can be very profitable work.

Chapter 17 – Murals

A new up and coming area of work for decorators. This has the potential for good earnings.

Chapter 18 – Architectural films

You have probably not heard of this but this is the future of decorating, especially in the commercial sector.

Contents

Section 5 – A bit arty

Chapter 19 – Decorative finishes

Graining, marbling and faux finishes. This is making a comeback.

Chapter 20 – Colour and design

Few decorators offer colour advice to customers, if you do then this can be a big advantage.

Chapter 21 – Window splashes

It is big in the States but not done over here, this could be a whole new trend that you start.

Chapter 22 – Pinstriping

Another one that earns good money in the States. You need to be good with a brush for this one!

Chapter 23 – Gilding

This is the application of metal leaf, usually gold.

Contents

Section 6 – Summing up

Chapter 24 – The last one

Here I discuss three things: common pitfalls that we all fall into as self-employed decorators, a few tips for building a bigger business and, finally, how we structure our time.

Section 1

Business pointers when specialising

In this section, we will look at some tips that are worth remembering if you are going to break away from the crowd and specialise.

Chapter 1
Introduction

Introduction

This book is aimed at decorators either thinking of leaving the firm they work for and setting up on their own or already have their own business but who feel they could do more.

When I wrote the pricing book, many decorators emailed me to tell me they had found the book useful. In the book, I chartered the journey of a young decorator who found his way as he ran his business.

I did this to put some problems relating to pricing into context and I used the story to illustrate points that I was trying to make. Many people enjoyed the business angle and have asked for more.

So here it is, more advice on decorating and having a business that you enjoy and that makes money.

When you work for a company you tend to do a certain type of work depending on your skills and experience. So, if you are good at papering then your company will get you doing more papering and leave the painting and the preparation to other painters who are good at that.

The company itself, however, may take on various types of work. They may do domestic work, commercial work, interior, exterior etc. They can do that easily because they employ several decorators, and each decorator has their own skill set.

When you leave a company and set up on your own, it is easy to think that you can do the same as your company did and offer a full range of services. You will offer interior and exterior. Domestic and commercial. You will put "No job too small" on the side of your van.

Introduction

The broader the range of work you offer, the more work is available to you. When you start up, you are looking to build up your work so you will take anything.

This is a mistake. You are only you. When you set up and start running your business, you are everything to the business. You are the person who prices, who chases invoices, who does the big outside job that you have decided to ladder.

You take on the wallpapering job even though you are a bit rusty and the last time you wallpapered was at college. You take on the big job that would have took your firm a couple of weeks but takes you 3 months.

You are run ragged, being pulled from pillar to post and not understanding why.

The best way to set up on your own is to transition from your job to your business. To illustrate what I mean let me explain what I did when I set up on my own and I will explain to you why I did it.

I had a good job at college that paid a good steady salary. However, I was not happy because I felt I could not do things my way. I went part-time at college (3 days) and set up on my own for the remaining 3 days. I had 1 day off, Sunday.

The money that college paid me for three days was enough to pay my bills. The business money would be extra. This took the pressure off when it came to taking on work. Now I know that I was lucky to be in that position, but you could do the same if you are currently employed.

I have a wide range of skills at my disposal. I can sign write, gild, wallpaper, paint and spray. I could have easily taken on a range

Introduction

of work. However, I specialised in one thing. I worked in a specific "niche".

That one thing was spraying. Not just general spraying either but airless spraying. I had my cards printed with "Airless spraying specialist" on them.

I knew that I could be more productive spraying and make more money, and I felt this would set me apart from everyone else because it was different.

It also meant that when I had job enquiries, I could filter them.

"Please could you come and wallpaper my hallway?" I was asked.

"No sorry, I am an airless spraying specialist, I can recommend someone, though," I would reply.

Then I had my first spraying enquiry. "I believe that you can spray out my house?" This was a builder who had refurbished his own house and wanted it spraying out, ceiling and walls.

I did the job (in a weekend) and he was not only over the moon but could not quite get over what I had done. He told everyone about the work.

Then I started getting, "Are you the guy that sprays?" enquiries and the work went from strength to strength. I was making good money and it got to where I was making more money spraying than I was at college, so I made the switch to full-time on the tools.

I also took on a similar type of work, generally spraying out full new build extensions or apartments. This meant that I was more

Introduction

or less doing the same work all the time, so I got good at it. I was spraying every day, so spraying was easy for me.

I was not switching from wallpaper mode to painting mode to outside mode to spraying mode. I could focus.

Customers got to know me for the work that I did and, eventually, I got more enquires for spraying than I could handle. I could choose what I was good at and work with the customers that I wanted to work for. Generally, the ones that would pay me what I asked and pay the invoice straight away.

This has been my experience and I feel that building your specialism is the way to go. You may not agree and if so then this book is not for you. I know many decorators do a range of work and they are happy doing that.

For many reasons, that I will explain in this book, I feel the way to start and build your decorating company is to specialise.

The difficult thing for you can be to think of a specialism. That can be a tough one, especially if you are just starting. I seem to be good at thinking up specialisms, so I thought it would make a great book if I listed several ideas and explored each one.

Introduction

Let us enter the world of the decorating specialism.

Chapter 2
Breaking away from the herd – choosing your specialism

Breaking away from the herd – how to do it

I have set the scene and if you are here in chapter 2, you have decided that specialising could be a great idea. Let us explore the idea a little further. What are the advantages of finding a decorating specialism and dominating it?

Many companies have a mission statement. This sums up what they are about. This is so that employees feel that they know what the company is all about.

Microsoft does not have a mission statement. There is an urban myth that one day some programmers were sat around a table in the canteen discussing the company. They wondered what the company mission statement would be if it had one. In unison, they all said, "Complete world domination".

Back when this story was told it looked like they might pull it off. These days we have Google, Amazon and Facebook going for the title.

Joking aside, it is a good thing if you can dominate a market and even for us decorators, I think it is possible to do that. However, to dominate the market in what you do, you need to specialise to the point where there is only you doing it.

This sounds odd but it can be done.

Let me explore a simple example.

Google "Painters and decorators" and your area. So, for me, it was "Painters and decorators Preston"

Hundreds of companies came up, many that I recognise. The first few had paid to be there. It costs about £4,000 to come number one for a Google search. It is a very crowded market. I am sure that your search will bring up very similar results.

Breaking away from the herd – how to do it

Now Google "Staircase finishers" & your town.

When I did this, no-one came up.

Now you could argue that is because no-one searches for that, but perhaps it's a brand new specialism that you have just created and once it takes off people will search for it and guess what? You will be the one everyone finds.

Back when I started as an airless spraying specialist, no-one knew what that was, but now if you Google it, you get quite a few people coming up in the search results.

It's not all about Google, even though everyone will tell you it is, it is about how people's minds work.

Let us try an experiment. Who makes the best chocolate?

Cadbury?

Who makes the best coffee? Costa Coffee?

Who makes your favourite fried chicken? KFC?

I could go on. You have two or three slots in your mind for the top brand in loads of different categories. Your top three favourite chocolates will be different to mine but often, the best in the sector occupy people's top slot in their mind.

To be the best painter and decorator in people's mind is a tough gig because there are loads of good guys out there. But to be the best airless spraying specialist for me was easy because there was only me.

Breaking away from the herd – how to do it

People would say, "Pete is the best airless spraying specialist". They even say that now because I was there first in my area. It is a great place to be business-wise.

It is sometimes hard to think of a specialism that you could dominate so, to make things easier, I have come up with as many specialisms as I could think of. I am sure there are more but these are my list based on my experiences in the industry and also on conversations that I have had with other decorators.

To dominate a specialism, you have to get really good at it and make sure people know what you do and how good you are. To get really good, you have to select a narrow enough specialism that you are doing it every day, so you get better and better at it, but not so narrow that you quickly get bored.

There are several ways to carve up our industry into smaller pieces. You can focus on a certain skill set or you can focus on a certain type of work or you could focus on a certain type of customer. Finally, you could mix the three.

Specialism number 1 – Wallpapering. You would only do wallpapering for a range of customers, both domestic and commercial.

Specialism number 2 – Houses over a million pounds – domestic only. You would build a customer base of wealthy people that needed the full range of decorating done.

Specialism number 3 – Retired professionals – You could focus on a certain type of customer that you are good at working with. Retired doctors and dentists may not have the inclination to do their own decorating, but they will expect a certain level of

Breaking away from the herd – how to do it

service if you deal with them. They will have friends who are doctors and dentists too.

Now we could take this a level further.

Specialism number 4 – Wallpapering for houses over a million pounds.

Specialism number 5 – Decorating for retired professionals who own house worth over a million pounds.

You can see that I have narrowed the specialism with number 4 and 5. I would narrow both the skill set you are offering, so that you get good at it, and the customer base that you deal with. They then see you as the "go-to" decorator for them.

I decided early on to specialise in spraying mainly in the commercial sector. Mainly apartments. I have since shifted my focus and I now do more freelance training and writing, which suits my skill set. You don't have to stick with the specialism you have chosen for the rest of your life.

Many companies start small in one small sector and then as they grow bigger, they expand their offering. This sometimes works (Virgin) and sometimes doesn't (Xerox).

You could do the same as you grow but you are better to focus first and then broaden out.

I was doing an online business seminar with three selected decorators all who wanted to build a better business. One chap was still working for a company and wanted to leave and set up on his own. Let's call him John Smith (not his real name, although I am sure he wouldn't mind me using this story).

Breaking away from the herd – how to do it

I was a bit mean, because I put him on the spot, but my intentions were good, and I just wanted to make him think.

I asked him two questions.

1. What are you going to call your business?
2. If I was going to get some decorating work done, and I approached you, why should I choose YOU and no-one else?

He was going to call his business "John Smith Painters and Decorators."

Now just to lay my cards on the table, I called my business "Pete Wilkinson Decorators" when I set up. "Fast and Flawless" came later. So, I have made all the same mistakes.

What does "John Smith Painters and Decorators" say to you? Let us look at the positives, everyone may know John Smith so people may say, "Oh yes, John is very good, he has been decorating all his life". People may also see the van and think, "Oh, I am looking for a painter and decorator. I may ask John for a quote."

Now let us look at the negatives. John Smith Painting and Decorating will get lost in a sea of painters and decorators. If someone he didn't already know decided to use him, he would probably have to compete on price as they are likely to have got more than one quote.

Also, everyone knows what a "Painter and decorator" charges. You could stretch and push for more but, on the whole, everyone has a rough rate in their mind for a "Painter and decorator".

Breaking away from the herd – how to do it

Finally, if John ever wanted to sell his business it would be harder because the business has his name in the title, it makes out that he is the main part of the business and it would not run without him. This is big and something that you do not think about when you start off.

I will discuss the name of your business in the next chapter.

The second question, why should I choose YOU and no-one else? This threw John and he mumbled something about being good at decorating, which is a given.

Many of us do not view what we offer from the customers' point of view. Let us look at an example.

You want to get someone to build some decking in your garden so that when it's sunny you have somewhere nice to sit with your beer and barbeque. You are rubbish at doing anything to do with wood, so it is definitely a job for a professional.

You have a few criteria. You want a good job doing, you do not want your garden to get damaged while the work is done, you want them to come when they say and you want to pay a fair price. You don't want it to take forever, either, and have the job drag on.

You get three quotes and here they are:

1. £600
2. £700
3. £750

Person 1 is not a tradesman but does odd jobs around the house and garden for people, he is cheap and that is his main selling point. You know someone who has used him, and his work was 5 out of 10.

Breaking away from the herd – how to do it

Person 2 is a joiner. He does all kinds of work, both domestic and commercial. He says he has given you a keen price because he makes good money on his commercial stuff, so he will fit your job in around that. You know he will make a good job but feel he may not make you a priority and the job might drag on.

Person 3 is a decking specialist who only works on domestic work, he has a good reputation and someone you know said that his work was 10 out of 10. He can come in the next two weeks and the work will take a day to complete. The work is also guaranteed.

Which would you choose? Yes, me as well, even though he is the most expensive.

Why?

Well, he specialises in what we are looking for. He does this work, day in day out, so he is slick. He is building a business doing decking so he will do a good job and turn up on time because it is his focus.

Okay, back to our decorating example. Why should a customer pick you?

This is how I would handle it. For a start, if the customer was not my type of customer, I would tell them: "I am sorry, but I am not taking any new clients". If the customer was my type of customer I would say: "This is exactly the kind of work that I specialise in. I have done a lot of jobs just like yours. I will give you a first-class finish and I will leave your house clean and tidy. I am honest and reliable, and I have references from previous customers."

Breaking away from the herd – how to do it

We will look at this in more detail in the next chapter. This is called a "pitch" and is important. It's also important that it does not sound like a rehearsed pitch.

Before we go on to the next chapter, I want to make something really clear. To have a good decorating business you need to be a good decorator. Obvious, eh? Well, sometimes people look for a shortcut and this may be why you are reading this book.

Maybe Pete has some ninja technique that will make me rich without having to put any effort in. Well, I am sorry to have to tell you this, no amount of marketing or positioning will make up for you being a rubbish decorator.

However, if you are a good decorator then this book will help you stand out from the crowd.

Another thing to think about is your chosen specialism. Pick something that you think you will enjoy. Odds are if you enjoy it then you will be good at it. For example, I can wallpaper, I am good at it, I used to teach it. Good money can be made wallpapering because a lot of decorators can't do it.

I hate wallpapering; it stresses me out. I would be no good specialising it in it, regardless of how much money I make.

Another thing we need to chat about is customers. I have spoken about this in previous books, but I want to recap here. Your job as a business owner is to build a raft of good customers. It's as simple as that.

Not cheap customers or pain in the arse customers; good ones. You are a good decorator, and you are in short supply and with the ageing population and the lazy millennial generation you will get even rarer. Your customers are not picking you; you are

Breaking away from the herd – how to do it

picking them. Good ones will give you a lifetime of income and be a joy to work for.

When a new customer rings me for a quote, I have several red flags that if they trigger one, I politely bin them off.

Here they are:

Is it not a spraying job? This is my specialism, and this is what I am after. Usually, they are, and that is why they are ringing me.

Where are they based? I do not want to go too far but will do most of the North West.

When do they want the job doing? If it's urgent then I likely will not be able to do it. I like customers who give me plenty of notice.

Then I ask about the actual job, what is involved. Then we chat about approximate prices before I go to have a look, this way if they have this idea that I will be £80 per day, I can save a trip to look at the job.

Finally, I hit them with the deposit, I usually charge 20% when they book the job. This has never been a problem but if it is then this saves me a trip to look at the job.

So, to sum up, don't do anything just for the money, focus on one thing that you enjoy and that you are good at and get known for it. Don't just work for anyone, choose your customers wisely.

Chapter 3
Your business name, tagline, and pitch

Your business name, your tagline, and your pitch

How important is the name of your business? How important is any name?

First, let us discuss the importance of names and the effect a name has on how people see you or your business.

An experiment was conducted by two psychology professors. They had a pile of written compositions for schoolteachers to mark. Some compositions had popular names on them such as David and Michael and some had unpopular names such as Hubert and Elmer (this was in America).

Amazingly, the teachers graded the papers with the popular names higher than the papers with the unpopular names!

Imagine how unfair that is? The teachers assumed that Hubert was a loser, and his paper would be of low quality and that is what the teacher saw when they read the paper.

What is going on here? Well, we are so busy these days that we see what we want to see. Our brain shortcuts the analysis process (because it's difficult and our brain can be a bit lazy) and it will jump to a conclusion with as little information as possible.

Another example is "Hog Island", which is in the Caribbean. No-one used to go there; it did not sound very appealing. Then they changed the name to "Paradise Island" and now it is the place to go for a luxury holiday.

Names are very important. Especially when naming your business. It's important for two reasons.

First, your customer will judge your business by its name. They will judge what you do, how much you charge and what you will be like to deal with. Amazing, really.

Your business name, your tagline, and your pitch

Second, when you come to sell your business, that you have built up to a nice profitable enterprise, the name you have chosen will affect the price you get for your business. The name could mean the business is worth nothing or £500,000. Yes, it could make that much difference!

This is a book about various decorating specialisms, so why am I going on about names? Well, you want your company name to tell the customer your specialism.

I saw a van on Homebase car park the other day. It was a VW Caddy and it looked very smart. On the side was a cool logo of a car and the company name "Just Dents" then underneath "Paintless Dent Removal", no telephone number, no big list of things that they do, no website address.

No need, because you can remember "Just Dents" and Google it.

JUST DENTS pdr
Paintless Dent Removal

Above is the logo that I have taken from the website. Now my next question is, "What do you think this company does?"

Your business name, your tagline, and your pitch

Easy eh? They fix dents on your car and they don't need to respray it. You get that from the name.

Do you think they will be good at it? Damn right they will because, guess what? That's all they do so they better be good, or they will go out of business. The branding on the van was so good I wanted to put a little dent in my car so I could get them to fix it.

Only joking of course.

The van was an expensive one and the colour of the van was classy. The branding was not fussy but well designed and thought out. He will have paid a professional to do the design. From all this, I assume that he will not be the cheapest, but he will be the best.

All from a glance of the van on the car park. This is how our minds work.

What has this got to do with you?

Once you have chosen your specialism you need to choose a company name that tells the customer your specialism. For each chapter, I will think of a name and look to see if the website is available. This is not so that you can use it (although if you're quick off the mark I suppose you could); it is so you can get ideas for yourself.

What if you already have a name for your business? Well, this is a good question because I called my decorating business "Pete Wilkinson Decorators". I know, I didn't know back then that this was a mistake. In my defence, I had a tagline underneath that said, "Airless spraying specialist".

Your business name, your tagline, and your pitch

Then as my business developed, I changed the name to "Fast and Flawless Decorating". The name "Fast and Flawless" told the customer what made me unique, I provided a flawless finish on their walls and woodwork. I kept the decorating in the title so that they knew that I was not a car sprayer.

Even though I advise people to remove the words "painting" and "decorating" from their business name, I didn't. You can break a few rules if you want to.

In fairness, if I started a new company, I would probably call it "Fast and Flawless Finishers" or "Fast and Flawless Interiors".

So, I change my business name from "Pete Wilkinson Decorators" to "Fast and Flawless Decorators". Is this a good idea? If your old name is no good, then yes, it is a good idea. Unless you're Coca Cola, no-one knows you so they will not notice the change. Even if they do, if it is a cooler name that describes your business then it could have a positive effect.

What if you have an established business that you have inherited off your dad? Everybody knows your name and the history behind the name has value.

No problem. Leave that business alone, it's doing fine. Set up a new business around your selected specialism and give that new business a cool name. Big business does this all the time. It gives the customer clarity. They don't even need to know that the businesses are linked but if they do then that's fine too.

Your van and any signboards that you use outside jobs need to be clean and clutter-free. I see so many vans with the name of the decorator in massive letters "John Jones" and then "painting

Your business name, your tagline, and your pitch

and decorating" much smaller underneath and then loads of text explaining what you do.

You drive past me, or I drive past your parked van. How much do I get from a glance?

"John Jones", that's what!

"Mmm, John Jones, I wonder what they do?" I think as I drive away.

I know your name is important to you, but I am afraid no-one else is interested. They are too busy thinking about themselves.

I saw a van with "Just Dents" on it and I knew what they did. Googled them and then I could read all about them.

No-one will read the side of your van, don't turn it into a page from a website. The world has moved on, everyone Googles it these days.

Your tagline

Once you have your name, you need a "tag line". In our example, the tagline was "Paintless dent removal". This is your opportunity to expand a little further what you are offering the customer.

Big companies do this as well; let's have a look at some of the more famous ones. See if you know the tagline before you turn the page.

Nike, Apple, L'Oréal, Ronseal and McDonald's.

Nike – "Just Do It"

Apple – "Think Differently"

Your business name, your tagline, and your pitch

L'Oréal – "Because you're worth it"

Ronseal – "It does exactly what it says on the tin"

McDonald's – "I'm loving it"

Wow, these are good taglines because we knew them already. They have gotten into your mind as a customer.

This is the game we are playing. You want to get a place in the customer's mind. It is very difficult these days because your mind is so full! It will reject as much as it can.

It will see "Dave Jones Painting and Decorating" and it will think "I already know a decorator, don't even bother remembering that", it will see "Just Dents" and think, "Oh hello, what's this? I don't know anyone who does this, I will remember that."

Okay, back to taglines. There are different types of taglines and we will look at a few but I like the descriptive tagline. A short sentence that describes what you do to the customer. So, mine was "Airless spraying specialist", not the best tagline, in fairness. "We spray your property to an awesome finish" may have been better.

Other types of taglines are:

1. A command

    These command the customer to do something, YouTube – "Broadcast yourself" for example. "Transform your property" might be one for us.

Your business name, your tagline, and your pitch

2. Thought-provoking

   Thought-provoking and stimulating. An example would be Adidas – "Impossible is nothing". For us, it may be "The finish is everything".

3. We are the best

   These are telling the customer that you are the best in your class. For example – Budweiser – "The king of beers", or BMW – "The ultimate driving machine". For us, it might be: "We take the paint finish to the highest level".

4. Ask a question

   This is where your tagline asks the customer a question. These can be good because the customer engages with your business by answering the question. A well-known example is Microsoft – "Where do you want to go today?"
   One for us may be: "Do you want your house to look amazing?"

5. Visionary

   An example of this would be Avis – "We try harder" or DeBeer's – "A diamond is forever". For us, it may be: "Spray finishing, the future of decorating".

You get the idea, descriptive is the easiest because the tagline is then doing its job of explaining what you do for the customer, but if you want to play around with ideas there are plenty to try

Your business name, your tagline, and your pitch

out here. For each specialism in this book, I will try to come up with a decent tagline for guidance.

The pitch

If you meet someone at a party and you get chatting about what you do then if you had little time you could just give your company name. "Oh, my company is called 'Just Dents'." "That's interesting," they reply, "what do you do?" "We do paintless dent removal." The conversation may move on but they are clear about what you do.

If you have more time, then you could use your pitch. Your pitch is longer and explains clearly what you do. This sounds cringeworthy. What? You're going to ask me to write down a script that explains what I do, learn it and then recite to everyone I meet?

Yep.

Make it sound like a conversation when you use it, don't read from a card! The more you do it, the slicker you will get. You can tailor it to the person you are talking to, so that it sounds spontaneous and like you're saying it for the first time.

When I was at college, I used to teach computer-aided signmaking. We had a cool machine that would take the design from the computer and cut it into vinyl. It was fast, and it was great to watch. I used to show it to people when we had an open day.

Did I just make it up as I went along every time we had an open day? No way, José. I had a set routine that I knew worked and got the "Wow!" response I wanted. I had a cool image already

Your business name, your tagline, and your pitch

set up and, once it was cut, I would give the person the finished item to take away. It worked like a dream every time.

When I was doing this routine, I would say things like "Oh, do you want to see the vinyl cutter working? It's really cool", if they said yes, I would say "Mmm, let's see what I have that would look really good", I would find my pre-set image, load the machine with a colour that I knew worked and away it went.

"It's amazing, isn't it" I would say. It all sounded very casual and made-up but it was a well-oiled routine that I had done hundreds of times.

Okay, back to our pitch.

Let us look at an example so that you get the idea. Let's say I was at a party and someone asked me what I do. "I have a training company and we teach decorators". Boring, eh? People would reply with, "Oh wow, that's interesting". Which means they are not interested.

Or :

"I have been a decorator all of my life and I have also taught decorators at my local college for over 20 years.

I found that decorators were not making use of spray technology in their business, so we have set up a training company to teach decorators how to get the best from the spray equipment.

One of our students, Jake, has gone on to double his profits in the six months after the course. He has bought some more equipment and his business is thriving."

Your business name, your tagline, and your pitch

Do you see the difference?

"Where do I find out more?" is the next question.

What did I include in my pitch?

1. What problem are you trying to solve?

Well, first, I told them the problem that we are trying to solve. There is no point in having a business that does not solve a problem. For me, the problem I saw was that decorators were not getting the best out of the available spray equipment, even worse they were not using sprayers at all!

2. Include your story

Your pitch needs to be part of your story so that the person you are speaking to can see where you have come from and where you are going. So, in my example, I make sure that they know that I am both an experienced decorator and an experienced teacher.

3. Explain the benefits of your business

Again, in my example, the benefit to the decorator of doing a course with me is that they could make more money on their future jobs. Double in the case of a real student that I use as an example.

Okay, you do not have a training company, you have a decorating business, so let us look at an example pitch for you.

"I have been a decorator for 20 years and I specialise in wallpaper. I have found that customers do not like seams in their wallpapered featured walls, so I install 'one-piece wallpaper' in the customer's property. My last customer, Alice,

Your business name, your tagline, and your pitch

has found that all of her friends have commented on the flawless mural and say that it makes the room look amazing."

I will devise an example pitch for each specialism in this book.

Chapter 4
Barriers to entry

Barriers to entry

If you have read my pricing book, you may remember me talking about "elasticity of supply".

What does this mean?

If the demand for decorators goes through the roof, then what will happen initially is that busy decorators will put their prices up.

What should happen then is that the higher prices will put people off a little and demand will level out. This is a short-term adjustment.

In the long-term, if prices stay high then people who are not decorators will think "mmm, there is good money being made here, I think I will set up a decorating business."

How easy it is for them to do this is what "elasticity of supply" is all about. If they can set up on Monday and be trading on Tuesday, then this means that the supply of decorators is very elastic. The market can adjust quickly, and prices will drop again.

Compare this to, say, a car manufacturing plant. This would take a while to set up. You would have to secure a building, specialist machinery and skilled labour. The task is complex and would need some serious finance to get going.

With our decorator example, someone may decide that they don't need any training, they buy a brush and a roller from B&Q and stick a pair of steps in the back of a car and away they go.

In the past, people with skills came together and formed guilds or societies to protect their commercial position. For example, I cannot just set up as a doctor. I need to be registered with the General Medical Council and, by law, I have to be qualified.

Barriers to entry

I would be a member of an association called the British Medical Association. They lobby the government to maintain "high standards" and by this, they mean hard to get qualifications. These alone create a barrier to entry so if doctor's salaries go over, say, £100,000 then I cannot set up in competition to bring the price down.

Solicitors and barristers do a very similar thing.

In medieval times, there used to be something called a guild that artisans and trades were a member of. The guilds made sure that people offering a trade gave a good service. You could not trade in the city unless you were a member of the guild and if you broke the rules you could be thrown out. Meaning that it would not be legal for you to trade.

All these things sound great and we all want to be protected a little from the market. However, a drawback is that it can make people become complacent and not improve their skills and productivity. There is a danger that they could get set in their ways because there is no incentive to improve.

It is very unlikely that the government will make it law for a decorator to be qualified. It is also very unlikely in this modern age that an association will become powerful enough to prevent people who are not members from trading as decorators.

I would forget all these shortcuts and dodges that create an artificial marketplace. We need to create our own barriers to entry.

How do we do this?

A good barrier to entry takes time to overcome. In our doctor example, it may take 5 years of training to become a doctor. In

Barriers to entry

our car manufacturing example, it may take a couple of years to set up a factory.

What barriers do we have at our disposal?

Our high level of skill

Despite what people say, to be a good decorator takes a high level of skill. When you look at the different trades, each has a mixture of knowledge and skill levels. An electrician, for example, has mainly knowledge, it takes little skill to run wires around the house.

A bricklayer has a high level of skill, it is very difficult to get as good and fast as an experienced brickie. Therefore, they make the big bucks when there is a house-building boom.

We as decorators have let our skill level slide over that last 30 years to the stage where anyone can pick up a brush and turn out some poor-quality painting. Because this is what people see professionals do, they don't know any better. However, if you look at some good painting then people know that they could not produce that level of finish.

Sharp cutting in, smooth flawless walls and perfect woodwork are achievable, but it takes time and patience to build up your level of skill.

Wallpaper is another area that decorators avoid because they cannot be bothered to get good enough. I would argue that the harder something is to master, the less likely someone will swoop in and steal your business.

Barriers to entry

Our branding

You may not think that your branding is a barrier to entry. In many businesses it is. The best example is Coca Cola. They spend a lot of money on the brand so that they can sell their cheap-to-make sugary drink. It is so powerful that when a rival tries to step in, Pepsi for example, people will argue that Coke tastes better, even if it doesn't.

That's the power of branding.

Pepsi cleverly did the "Pepsi challenge" so that they showed that many people actually preferred the taste of Pepsi. However, it makes no difference, Coke is still number 1.

Common drugs are an interesting one. When a drug company develops a new drug, they have a patent on it. They can sell it at a premium under a brand name. Once the patent expires, the drug can be sold by anyone at any price. Because it is a drug, it is made to a very strict specification.

An example of this is paracetamol. These days you can go into Asda and buy paracetamol for a few pence. However, there are still brands out there (Panadol for example) who charge much more for the same product.

I watched a programme on the television about this and some people had it explained to them that the paracetamol was made to the same formula, regardless of the branding.

They were then offered two packets of tablets, one branded for £1.80 and one unbranded for 50p and quite a few still went for the more expensive branded one because that is what they had always bought.

Barriers to entry

Amazing!

That is the power of brands in people's mind. Therefore, you need to develop your brand, made up of your name, logo and tagline.

The equipment we use

When I talk to decorators about spraying, the conversation quickly comes around to the cost of the equipment. A typical professional mid-sized airless sprayer is about £2,000. Decorators initially are shocked at the cost and see this as a bad thing.

But it isn't, it's a good thing.

Why? Well, Dave down the pub who will paint your lounge for £100 will never go out and spend £2,000 on kit. If you build a band of loyal customers who want you to spray their work then for someone to compete with you they will have to go out and spend £2,000, which they are less likely to do.

Dustless sanding is another one. No customer likes dust in their house and no decorator likes breathing it in. Dustless sanding is essential these days if you are a professional decorator. They are not cheap, and you can easily spend £1,000 getting set up. Sometimes more, depending on what you get and how many decorators you employ.

Again, this is another barrier to entry for your casual "have a go" decorator.

Barriers to entry

Our knowledge of products and systems

Hundreds of different surfaces need to be painted: plaster, brick, block, softwood, hardwood, copper, steel, plastic, uPVC, composite, fibreglass, aluminium and much more, the list is constantly growing as new materials are developed. Each surface has its own considerations for applying paint and you need to use the correct primer.

Then you have paints. There are primers and topcoats, there are varnishes and lacquers, there are emulsions and masonry paints. We have different sheen levels from eggshell at around 10% sheen through to gloss. Some companies will even make a custom sheen level just for you, 37% sheen for example.

Some paints have been out for years and new paints are being developed all the time. If you look at the number of combinations of paint, surface and finish, there are millions, and more being created every day. Even someone who works in the industry has difficulty keeping up, let alone a newcomer.

This is a big entry barrier, and it pays to stay up to date with products and systems as much as you can. If a customer asks you a question and you can confidently answer it off the top of your head, then that goes a long way to building their confidence in you. You can still go away and verify your recommendation with the supplier.

Our loyal customer base

I have put this last because many decorators put too much importance on this factor. They assume that all their customers will stick by them forever. These days that is not the case.

Barriers to entry

You can still look to build a loyal customer base and work hard to keep your customers happy.

Your specialism

Choosing and getting very good at your chosen specialism is another entry barrier. The better you get at your chosen skill the faster you will be, and the work will be of a higher standard. This makes it harder for someone to set up and compete against you.

Section 2

The basics

In this section, we will look at some bread and butter decorating markets, nothing specialist, just to set the scene and show you how a specialism would work for you.

Chapter 5
Domestic interiors

Domestic interiors

I thought I would start simple with the work that we all do as decorators. The advantage of domestic work (as opposed to commercial) is that it tends to be of a higher standard and can be more enjoyable. You have more control over the process and, generally, you are also dealing directly with the customer.

Above - A cool lounge

Another advantage is that you can handle a larger number of smaller (by job value) customers so that you are spreading your risk. If you have 200 customers and you lose one, it is not the end of the world. However, if you get all your work from one main customer then you are very vulnerable if they decide to use someone else.

Generally, you will find that once you have built up several private clients then you are at full capacity. To expand, you would look at taking on new staff or you may decide not to take on any new clients unless you lose one.

Domestic interiors

I like doing interior work because it is not weather dependent, I can schedule jobs back to back knowing that I will not be rained off, which creates a backlog of work.

Because domestic interiors are one of the main things that all decorators do, you must be a little bit creative to make yourself stand out from the crowd as a specialist. You cannot just say "I am a decorator who just does domestic interiors" that will not differentiate you.

The idea, and we will explore this with every specialism that we look at, is to build a "value proposition" for the customer. This way you are focussing on what value you are giving to the customer.

Now I know what you're thinking, you are thinking "what is Pete on about now? I am decorating their lounge, for God's sake, that's the value I am giving!"

Yes, I know that, but that is also what every other decorator that competes against you does. Let us look at what a typical decorator gives a customer.

1. Turn up on time.

2. Protect the furniture and carpet.

3. Provide the materials.

4. Decorate the room to a high standard and take about a week.

5. Tidy up.

These are things that a customer would expect you to provide. What could you also provide? This is more difficult to answer and the best way to find this out is to ask customers what things they would like you to do for them.

Here are examples of what a typical decorator would not do.

1. Clear the room for the customer, much like a house removal company does. You could carefully package up their belongings in bubble wrap and put them in storage cases.

2. Spray the woodwork to a "car-like" finish.

3. Provide daily updates on the progress of the work.

4. Complete the job quicker than a normal decorator, maybe 2 days instead of 5 days. You could achieve this in several ways, but you are minimising disruption to the customer.

5. Clean any surfaces not being painted: sockets, light switches, plastic windows, etc. This way everything looks new.

6. Clean the carpet while the room is empty, either professionally with another company (this would be best) or get the equipment yourself.

These are just a few ideas that I have thought of while sat here writing, I am sure that you could come up with more. The advantage here is that the customer sees you differently.

They would say things to their friends like "Look at the woodwork. Isn't it amazing?" or " he cleaned the carpet as part of the job" or " they were in and out in a couple of days" or "they cleared the room for me and then put everything back the way it was".

All these things remove hassle from the customer's life and that is why they are paying you. We have all known decorators that

Domestic interiors

leave the room in a poor state. They feel that they have painted, so that's them done.

In the States, decorators have a different approach to us when redecorating the inside of a house.

They do this.

1. They ask the owners to leave the house and stay in a hotel.

2. They clear the whole house and pack everything away.

3. They paint the whole house quickly, in 5 days.

4. They finish the woodwork to a high standard.

5. They charge a good price for this service.

What are the advantages for the customer of this approach?

They are not in the house when it is dusty, smelly and messy. They do the whole house so the whole house looks fresh. The problem with the one-room-at-a-time approach is that it leaves all the other rooms looking drab.

It is also cheaper for the customer because it is more efficient.

What are the advantages to us?

It is easier to do the whole house with no-one in, you need not leave it ready to use again every night. We make more money doing a full house but can charge less than we would if we did the job room by room because it is more efficient.

You can use more productive methods and the customer is not there to time how long you spend and then reverse engineer the price.

Domestic interiors

Everyone wins.

Another advantage of this approach is that no-one does this in the UK so that makes you unique. You know the method works because it is done over the pond.

Here we have specialised using a unique approach. Another way to specialise in interiors is to choose a particular type of customer or price bracket. Let us explore the price bracket first.

Don't forget: with the specialisms, you are looking for gaps in the market that no-one else is looking after. Price-wise there are two obvious gaps, although in fairness decorators do try these two approaches already.

The high price approach.

Here you must remember that some customers want to pay more. Take cars, for example. We could all drive around in a £1,000 second-hand car. It would do the job and it would be cheap motoring. However, some people will buy a Porsche.

These are just over £80K.

Why do some people pay so much for a car? The main reason is exclusivity. Few can afford a car that costs this much so owning one shows that you can afford it.

If you go down this road, make a name for yourself as an exclusive decorator that is hard to get hold of. You need to provide an amazing service that blows everyone else out of the water. You cannot ALSO do "low-end stuff" otherwise it destroys the illusion.

Domestic interiors

This is a tough specialism to get into and to be honest I would not bother because there are easier ones to attack as we will see in the later chapters. It is, however, a strategy that some use.

Above - A simply decorated bedroom

The low price approach.

Amazingly, this can give you more profit because it is so much more scalable. It is an approach that many take when they start up and try and get work.

For this to work, you need to get customers who are not that worried about the finish but just want a clean-up job doing and you need to deliver it quickly so that you make money.

You also need to scale up big to make serious money.

I have started with a specialism that is difficult but a good one to explore with new ways of thinking. Some of the later specialisms

Domestic interiors

are easier to set up and require less creativity simply because they are new to the market.

Let us explore the name, tagline and pitch.

Name: Just Interiors

Website: www.justinteriors.co.uk

Tagline: A full house decorated in a week.

Pitch: We have been decorating for years and we found that what customers want is a full house redecorating at a fair price and not be there while the work is done. Our last customer went on holiday and when they returned the whole house was freshly redecorated and everything was back as they had left it. They were over the moon.

Chapter 6
Domestic exteriors

Domestic exteriors

I know that I have already said this, but I don't do exteriors. Why am I saying it again? Well, because a lot of decorators say the same. In this country, we have a problem, and that is the weather. It not only rains a lot, but it is so unpredictable!

It can be sunny at 8 am, raining at 11 am and sunny again at 3 pm. It can be sunny at your house and raining on the job. Or even worse, sometimes the other way round. You have the customer on the phone wanting to know where you are and when you say "It's raining" they don't believe you because it is not raining there.

Because so many decorators avoid externals, this means that the ones who do them can make good money. As with all the specialisms, you must get good at them.

In previous books, I talk about Chris Berry. an American painter who has a company called B&K painting. He mainly does outside work. He is very good at it. He has a team of four guys, and they turn around one outside per day on average, that is five a week.

Now I know that is just an average and some outsides will be bigger, and some will be smaller, but this approach gives Chris a good turnover and profit.

He has the weather on his side, they have sunny days most days through the summer months and this is a big advantage.

I think you could replicate this business in the UK. You would need to put systems in place to get more efficient at painting outsides and I have discussed these in a previous book called "Fast and Flawless Systems".

Domestic exteriors

I think if you could solve the weather problem and find some way of keeping the elevation that you are working on dry then I think that would put you ahead of the crowd.

How could this be done? Well, I can think of a couple of ways, but I will let you ponder the problem for yourself.

Specialising in exteriors is not a new specialism for painters and so I have got it in the "basic" section. It's an easy-to-understand specialism that can be very profitable if you can manage the weather.

I know a local decorating firm with a team of five decorators who do mainly exteriors. He will take on any size of exterior and has a deal with a company to hire cherry pickers and scissor lifts. He seems not to be bothered with the weather and always seems able to work on the "dry side".

I did some searching on the internet and found one or two American companies that specialise in painting the exterior of people's homes in one day, with a team of painters using systems and products that speed up the process. I am not sure how well received this would be in the UK however it would be unique.

Domestic exteriors

Company name: Simply Painting Exteriors

Website: www.simplypaintingexteriors.co.uk

Tagline: Exteriors painted with our weatherproof systems.

Pitch: We have been painting exteriors for many years in the UK and we have found that the biggest problem is managing the weather. We have developed a system where we weatherproof the property while we work so that the work is carried out to specification and on time. Many customers comment that it was difficult to find a company that would do the work in the required time frame and that we exceeded all expectations.

Chapter 7
Paperhanging

Paper hanging

I have been decorating for nearly 40 years and over that time wallpapering has come into fashion and gone out of fashion and then come back in again.

In the 1980s, people would wallpaper everything, every wall in the lounge, the whole of the hallway landing and stairs. This was good for decorators because most people would not tackle wallpapering the stairs, it was just too tricky.

When I served my time, my company wallpapered a lot to the extent where I got to wallpaper even as an apprentice. Along with the practice I got at college I was confident at wallpapering.

When I went self-employed (the first time) I did a lot of wallpapering jobs. Back then it was expected that, if you were a decorator, you could wallpaper.

Then decorative techniques came into fashion in the 1990s; it was all rag rolling and sponge stippling. The people that jumped on board this train made some good money because at the time not many were confident with those skills.

We ran courses at college at the time and we had hundreds of people doing the courses, mainly due to the "marketing" of television programmes like "Changing Rooms", which was an interior design programme where designers/decorators went into people's homes and marbled them and the like.

This lasted quite a while, most of the way through the nineties.

Then it was all plain, plain, plain. White ceilings and walls and woodwork. No wallpaper, no decorative techniques. Plain.

So, you can see for quite a big chunk of time no-one has had wallpaper done except maybe the odd granny stuck in the past.

Paper hanging

Then, more recently people have been introducing a bit of colour in their lives, well, grey anyway, and also having a feature wall in wallpaper.

Back in the eighties, a roll of wallpaper cost about £6 so it was no hassle if you ruined a roll because you cut all the lengths too short. But now, people are choosing £150 a roll wallpaper for their feature walls.

Why wallpaper is so expensive now I do not know, the only thing I can think is that it has been out of fashion for so long that companies have stopped making it. So, supply is short.

So, we have a perfect storm of decorators who have not hung wallpaper for nearly 30 years, some decorators have never hung wallpaper in their career. Anyone willing to give it a go has to risk messing up £150 roll wallpaper and they have the liability of paying for new wallpaper if it all goes horribly wrong.

This does two things, it put decorators off from taking on jobs, thus creating a shortage, and it drives the price up for the decorators that can do it well and are confident to hang wallpaper.

It would not surprise me if people started asking for the whole room to be wallpapered, too, as they did back in the eighties.

What does all this mean?

It means that it could make a good specialism for any decorator that likes wallpapering and wants to get up to speed. You could offer a specialist wallpaper service to customers and you could also subcontract to other decorators that don't want to do it themselves.

Paper hanging

There is a healthy market for this, and I think you can earn some good money doing it. I have heard of various rates being bandied about from £15 a roll to £40 a roll. For people that charge by the metre, I have heard of rates from £5 metre to £15 per metre. The most I have heard is an eye-watering £56 per metre.

Over in the States, the skill of wallpapering is even rarer. They do not price per square metre they price per square foot. A square foot is ten times smaller than a square metre. So, if you charge £50 per square metre then that is only £5 per square foot.

I was discussing rates for murals with some guys over there and when I quoted a figure of £5 per square foot, they thought it was too cheap! Maybe learn how to paper well and get yourself over to America. It will vary from state to state. Paperhanging is seen as a different trade over there, separate from painting.

Anyway, back to earth. It is worth exploring this as a specialism for yourself if you have a flair for wallpapering. There are various markets out there, you could pick one or you could go for a few to spread the type of work that you do.

There is the domestic market: hallway landing and stairs, lounges and bedrooms, that kind of thing. You could do some commercial: hotels, restaurants, pubs and offices. There is quite a bit out there. Commercial work tends to be wide vinyl so you would have to get up to speed on that. There are murals that you could offer too but I am going dedicate a whole chapter to that because I think it is such a good specialism.

Many builders will get in a separate firm to do the wallpapering because many site painters are just that, painters. I think it may

Paper hanging

go the same way as America and Australia and become a separate trade. A good specialism to consider, I think. It's clean and tidy and you can get good photos and visuals for your marketing.

It beats a photo of a plain white wall!

Paper hanging

Company name: Wallpaper Wizards

Website: www.wallpaperwizards.co.uk

Tagline: Wallpaper perfection everytime

Pitch: As a company, we have been wallpapering for many years. We have hung paper ranging from lining paper to grasscloth. We find that 95% of decorators are not confident with hanging wallpaper and therefore we fill that void. Many of our customers are decorating businesses themselves and they get us in to do the specialist work.

Chapter 8
Fences – Commercial or domestic

Fences – commercial or domestic

This is a specific specialism. It is something that anyone could do. However, there are two things about painting fences that you need to remember.

First, no-one likes doing the job themselves and every DIY person who does it are slow and make a mess when doing it.

Second, there are thousands of miles of fences up and down the country that need painting or staining to make them look good.

When I went part-time from college, this was the business I was going to set up. My plan was simple, you can spray a fence panel in 30 seconds, 1 minute if you do both sides. If you set up a

Fences – commercial or domestic

gazebo you could have a rainproof spray area where you could work.

My main selling point was a transparent pricing policy so that people could see how much the job would be. For example, you could charge £5 per side for a coat of stain. An extra coat would be an extra £5. So, if they had both sides of one fence panel done with two coats it would be £20 per panel.

You would have a minimum charge so that no-one tried to get you out to spray two fence panels. You could also give some discount for bulk so that someone with 100 fence panels would get it cheaper.

The advantage was that people would be painting their fence on a Sunday and it would take them all day just to do one side. They would then look at the prices and think "for a fiver I could have had this panel done for me including paint."

Fences – commercial or domestic

If the typical garden had 20 fence panels and they had 2 coats and they also did both sides then that would be £5 + £5 (2 coats) X 2 (both sides) X 20 = £400

You could easily do this in a morning with a sprayer. Fence paint is cheap so on an ideal job like the one above you could make some good money. On commercial work, you could easily charge those prices and probably get the contract to redo the fences every 3 or 4 years so you have repeat work.

Another advantage of this approach is that it would be easy to scale to a larger area, you could train people up to deliver the service as you developed a growing workload. You could pay your staff on a per panel basis so that they made good money and so did you.

A disadvantage is that it would be easy for someone to move into the market and set up in competition if they saw that you were doing well. The trick would have to be a slick service that could be done in any weather and create little mess or fuss.

A partnership with a fencing company would work well as you could paint the new fences that the company installed. If you got this right, you could paint the fences before they go out on-site to be fixed. You could do a good deal for the fencing company so that they could charge the customer extra to have the fences painted and make a percentage from that without actually doing anything except giving you the work.

You could expand the business to include decking and sheds and all things garden and outdoors related. If you are a qualified decorator, then this will tip the balance in the customer's mind if you were competing against a handyman as long as you were not too far out on the price. It would be important to deliver a

Fences – commercial or domestic

first-class result quickly so that the customer is happy, and you make some money.

Fences – commercial or domestic

Company name: Timber Treatments Ltd

Website: www.timbertreatments.co.uk

Tagline: Take the hassle out of painting your fence.

Pitch: We have found that to paint a fence correctly can take a do-it-yourself homeowner hours and hours of your precious weekend or evenings. Our service is quick and professional giving you a first-class job for a reasonable cost. All our customers have found that it saves them time and hassle getting us to keep their outdoor timber in a perfect condition.

Chapter 9
Churches

Churches

I have a lot of experience in painting churches. The company where I served my time did a lot of churches and because I was the apprentice signwriter I tended to work on them all.

We were good at doing the church work, the same team of decorators did them and we were fast at getting the work done. A few things make church painting specialist.

These are:

The height

Most churches are high and have a peaked beamed ceiling, this makes access difficult. There are several approaches: these days you would have to scaffold the church out so that you could work on the ceiling. You could use a cherry picker if there was enough floor space. Back then we mixed mobile towers for the lower ceilings and scaffold for the high ones.

Above – An old photo of a church I decorated and gilded

Churches

The specialist skills

Another aspect of church work is the specialist elements of the decorating. In fairness, when we did the churches it was mainly just matt emulsion on the ceilings and walls. Maybe varnish on the beams. The majority of the work was carried out by normal decorators. There are elements of the work that is specialist, there may be some stencil work, or signwriting or artwork.

Our company had a signwriter for the run-of-the-mill artwork and a freelance artist who came in and did the specialist stuff.

Because of the specialist nature of a lot of the work, decorators shy away from doing this job. Many self-employed decorators will feel that the job is just too big for them to take on.

When I was self-employed the first time around, I took on a couple of church jobs. I was experienced at doing them at my old firm and I had the specialist skills needed to carry out the work.

I got a company to scaffold where needed and I used a couple of sub-contractors to help me out with the bulk of the straightforward work. Because few companies out there do the work, it was not very competitive at the pricing stage either.

Churches do not get decorated often as it is an expensive job, so it is not a regular repeat business model. However, our firm almost always had a church job on for many years. There are a number of churches in every town and it's surprising how much work this adds up to.

Once you get into doing them you will find that your name will be passed on so that work will come to you. We did mainly

Churches

Catholic churches, for two reasons, my old boss was a Catholic and they had enough money to pay to have the work done.

Just out of interest, I Googled this to get some interesting facts and I found that the Catholic Church is the richest organisation in the world, worth between $400 billion to $2 trillion. Just remember that when you put your price in.

To specialise in church work, I think that you would need to get some experience of this kind of work, maybe subcontract to a company that does it already. It would also be worth getting some specialist skills or at least finding someone that could help you out.

As you can see above, not all churches are full of fancy artwork and gilding, some of the more modern ones are plain and straightforward to paint.

Churches

I think it would be difficult to break into the inner circle of people that decide who paints the churches but if you did a bit of legwork and spoke to priests and vicars of your local churches you will get an idea of who the current players are and how much work is available.

Long-term, if you built a customer base of church work, I think it would be hard for other decorators to muscle into your market, especially if you had a good reputation and you were very good at it.

Above - Here is a church being painted, photo courtesy of John Kerry

Churches

Company name: Decorators for the Church

Website: www.decoratorsforthechurch.co.uk

Tagline: We take the hassle out of decorating your church

Pitch: We know how difficult it is to find a decorator that can handle all aspects of church decorating, there needs to be consideration of the use of the church while the work goes on and the work needs to be carried out safely and to a high standard. Many of our customers use our services because we can carry out all the specialist work needed and we handle the access arrangements safely and with care for the fabric of the historic building,

Section 3

Spraying

Chapter 10
Spraying overview

Spraying overview

Spraying is my specialism. This means a couple of things: I know a lot about it, and I have written a lot about it. I considered not even including it here but then I decided that this might be the first book of mine that you have read and it would be a shame to deprive you of one of the best specialisms out there.

The good thing about spraying is that it is a massive specialism and these days almost a separate trade. A few specialisms in this book are spray related so I will not talk about everything here.

I will try to summarise spraying and look at the different spray systems out there and how they fit into the grand scheme of things.

One thing that I have noticed is that there is a wealth of information out there on all things spray related. There are Facebook groups, YouTube channels and websites.

Unfortunately, if you know nothing about spraying you can be left more confused than when you started.

Another thing is that because the internet is global you get differences with different countries. I see lots of examples of conversations that go on between two decorators from different countries and they are arguing about something without realising that they are both right, it's just that the circumstances are different in their country.

A concrete example of this is back rolling when spraying. If you are spraying a textured surface such as Artex or blockwork then you need to back roll it to get the paint to penetrate the surface.

In America, most walls are textured, they call it a popcorn finish or knockdown. In the UK, walls are smooth. But Americans don't know that our walls are smooth, and we don't know that their

Spraying overview

walls are textured. So, when a discussion about how to paint walls crops up, we are immediately at cross purposes without realising.

There are many examples of this so remember this when researching spraying.

Why did I choose spraying as a specialism? Well, there are several reasons:

1. **Few people are doing it.** Even now there are still towns that don't have any people that are sprayers. In Preston, where I live, there are still only a handful of decorators that make most of their money spraying.

2. **It costs quite a bit to get into it.** I have discussed entry barriers already, but it is one reason I like it. Someone would have to spend a few grand before they had the equipment that I have, so this puts them off competing.

3. **The finish is amazing.** It is easy to sell the finish to a customer. It makes you different from all other decorators and it is something that people will pay for.

4. **It's fast.** This means that you make more money in less time.

5. **It's physically easier.** This means that you need not work as hard to get the same results as a "standard" decorator.

6. **It's fun.** This is a hidden benefit. It's fun going to work when you are spraying.

I could go on, there are many reasons that I like the specialism that I have chosen for myself. Let's have a look at the types of

Spraying overview

sprayers that are out there and discuss how much they are and what you would use them for.

Conventional

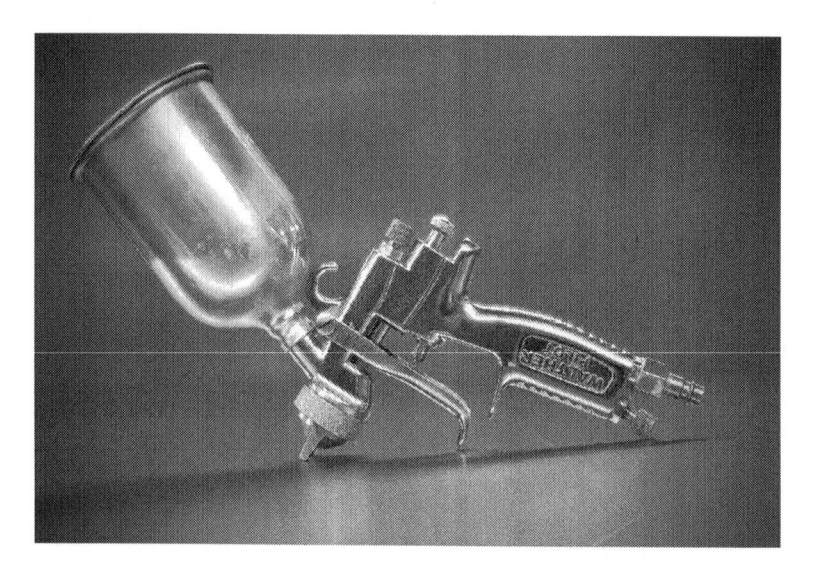

A conventional spray gun works off a compressor and looks like the photo shown above. The advantage of this system is that it is cheap to get set up. You can buy a compressor for a couple of hundred quid and a gun for a hundred and you are away.

Once you get your head around the controls on the gun, you can use this system to spray many smaller areas such as doors, windows, radiators and skirtings. You will need to use the right paints to get a good result.

High Volume Low Pressure (HVLP)

HVLP systems are very similar to a conventional spray system but, instead of a compressor, it uses a turbine. This produces the air that atomises the paint. The advantage of an HVLP system is that it is small and very portable. This disadvantage is that they

Spraying overview

are more expensive than a conventional setup. Typically, a thousand pounds.

An HVLP system, small and portable.

HVLP would be used in smaller items and woodwork just the same as a conventional setup.

Spraying overview

Airless

This is a powerful pump that pumps the paint to the gun. In the gun, there is a tip which has a very small hole (around 17 thousandths of an inch) and this turns the paint into a spray.

It is fast and is ideal for painting large areas such as walls and ceilings.

A small airless system

They can be tricky to master and there are dangers involved with the high pressure, so you need to do your homework before using one.

Spraying overview

Air Assisted Airless (AAA or Aircoat)

This is a mixture of airless (giving you the power to atomise thicker materials) and conventional that gives you air to soften off the paint stream.

These are expensive and cumbersome systems and are better suited for a workshop/spray booth set up. Typically used by joinery manufacturers. They can be set up on-site, but this is rare.

Above – An air assisted airless system

Spraying overview

Notice the two hoses to the gun, one for paint and one for air. The pump itself would be attached to a compressor for a source of compressed air.

This gives you an overview of the spraying systems out there. I would first decide what I was going to specialise in (more on specific spraying specialisms later) and then buy the system best suited for that type of work.

Once you have chosen a system, let's say HVLP, then buy the best HVLP that you can. If you cannot afford it then save up. Do not be tempted to buy a cheap one. You may drop lucky and find a good one second-hand.

There is much more information about airless spraying in my first book, "Fast and Flawless – a guide to airless spraying".

Chapter 11
Apartments

Apartments

My first real spraying job was some student apartments in Salford. I had set up on my own as a decorator and I had my own airless sprayer. I have a good friend who has his own decorating business. He has been self-employed since 1990 so he has been at it a while!

We were sat in a pub having a chat about work, as you do, and he mentioned that he had priced a job for 80 student apartments. It was a redecoration, the walls were magnolia, the ceilings were white, and the woodwork was white.

They were being redone all white for the walls and ceilings in durable matt. An additional bonus was that the carpets were being removed to be replaced and all the furniture was being removed to be replaced with new furniture after the decorating was done.

It was like a gift from God.

"You have GOT to spray those," I said.

He was not too sure, he muttered something about not being able to touch up and went off to buy another round.

To cut a long story short, we sprayed them, and it was very successful. We had a bit of learning to do to get the system right but once we got going the progress was fast.

We got so good at it that one year he was asked to drop his price. He refused, and another company did the work. They failed to deliver, and we were back "on it" getting the job across the line.

Apartments offer you quite a few advantages when it comes to spraying. First, you have the volume so you can set your stall out

Apartments

and work from room to room very fast. I was lucky in that the builder was very organised and I had a clear run of two floors at any one time. No other trades got in my way for the short period that I was spraying out the apartments.

We had the masking system down to a "T" and it did not take long at all to mask a floor of 14 apartments.

I did several apartment jobs over the years and in many ways (before the training academy) it has become my specialism.

There are several advantages to specialising in apartments and one reason that it is my favourite type of work.

First, there is plenty of work around. There are constantly new apartments being built and old ones being refurbished. This means that if you want there is a steady stream of interior work for you to do. You can decide how many days you want to work and have a nice steady business.

Second, if you can get your systems right then they can pay well. It's not easy negotiating the price or landing the job but once you do (as long as you have not priced too cheaply) then you can make good money.

Third, at the moment (2020) the majority of apartment work is brushed and rolled. Even though it's quicker and better to spray them, there is resistance to it. This means that if you can convince the client to spray (this will get easier and easier) then you can double your productivity compared to the brush and roller guys.

I have written a lot about my apartment work in all of my previous books so I will not go on too much here. However, it is something to think about.

Apartments

Company name: Apartment Refinishers

Website: www.apartmentfinishers.co.uk

Tagline: The next level apartment finish

Pitch: We have been painting apartments for many years and what we have found is that very few decorating companies can get the work over the line in the allocated time and get the quality to the standard expected. One builder commented on one of our show apartments: "this is the best painting that I have seen in my life". Enough said.

Chapter 12
Commercial shop shutters

Commercial shop shutters

This is interesting because I discovered it by accident. I rarely do exterior work, but we were working on a job that we do every year and the chap asked us if we could spray his shop shutters.

There were four large shutters and four doors to do. This was a straightforward job and I power washed them down, prepped them, masked them and sprayed them comfortably in 2 days.

This was the first one that I had ever done so I feel I would get faster with experience. I didn't charge enough for this job, but I found that you could charge £250 per large shutter and £150 per door shutter easily. This would be £1,600.

Above – the first one I have ever done

Commercial shop shutters

Above – The preparation

Commercial shop shutters

Above – Undercoat

Commercial shop shutters

Above – The finish - gloss

Once I had done this, and it looked great I was approached by another shop owner for me to do his shutters. This job was smaller with two large shutters and two doors. For this, I also charged the same as the last job even though it was smaller (I was getting wise to the pricing system) and he happily accepted.

Commercial shop shutters

This job was completed in about the same time because I took my time and did a good job. This one looked great too.

Then I was approached by another owner. Remember that I don't do outside work, I had done no marketing, it was simply word of mouth and shop owners who wanted to freshen up their shutters. I think my initial "cheap" price also had some influence.

I priced the third job and got the work. Once I had completed this job, the weather was turning, and I was approached by another shop owner. This time I decided that this was not the specialism for me due to the weather constraints and I declined the job.

He was persistent and it took a while before he finally took "no" for an answer. I was amazed by how much demand there was out there for this type of service.

I have discussed this idea with one or two decorators, and I get a similar response, there are just not enough shop shutters out there to keep me busy.

Let us think about that for a moment.

First, how many do you need? I did three jobs and took approximately £3,000 with just over a week's work. Depending on how much you want to earn a year you could do three jobs a month, so that's 36 a year and you would earn a reasonable amount with more time off. You could even focus on the work in the summer leaving the winter free.

There are over 300,000 retail outlets in the UK, so if you got 1% of those that would be 3,000 shutters. At a good price of, say, £1,500 per job that would be £4.5 million. Phew! Okay, let's say

Commercial shop shutters

we only got 10% of those 3,000 shutters, that would be 300 jobs.

300 times £1,500 is £450,000

As you can see, you don't need that many to build a nice specialism business. I think if you did specialise in shutters alone you would have to travel, possibly even nationwide. You could expand your specialism slightly and also do cladding and other elements of the retail outlet.

What about online shopping and the decline of the retail outlet? This is a good question, and I am not an expert on the future (if I was, I would be a millionaire). However, note that one of the biggest online retailers, Amazon, is buying up physical stores in the States and they are looking to set up cashier-less shops.

If Amazon is doing this, then you can bet your bottom dollar they think there is a future in it. Do not underestimate that there are still physical buildings even with an online business and these buildings still need to be painted.

This is a good specialism and one that you could expand and get repeat customers. If you did a good job, then shops return to redo the shutters when they need doing. You could even integrate a maintenance system where you power wash the shutters and touch up mid-cycle to keep them looking fresh.

What paints would you use for this kind of work? Well, I won't specify the work for you, that is your job. If you are going to specialise then you need to get your head around the latest and best products on the market and test them.

Commercial shop shutters

You could contact paint manufacturers and get them to specify the jobs as you get them. You will soon get a system that you like and is cost-effective.

I will give you some pointers.

You could use exterior undercoat and gloss. Just a standard system. You could even use an oil-based system. However, I would check out Rust-Oleum. They make a range of products suitable for this kind of work, a popular one is Noxyde.

I would get a Rust-Oleum representative down to your job and ask them to specify the work. The Noxyde product is very good and it is a fair price too. There are other similar products on the market, but I will let you find those for yourself.

Commercial shop shutters

Do not cut corners on the product that you use. You want the shutters to last and you want the customer to keep coming back to you every time the work needs doing.

For the work, I used an airless sprayer (see the chapter on spraying) however you could use a simple conventional system to do this kind of work if you wanted to.

Commercial shop shutters

Company name: Shutter Painting Solutions

Website: www.shutterpaintingsolutions.co.uk

Tagline: Shop shutters transformed from shabby to pristine.

Pitch: We have been renovating the outside of properties for many years and we have found that it is very difficult to find a contractor who can repaint your shop shutters so that they look like new while the shop is still in use. One shop owner commented that he could not believe the transformation that painting the shutters made to the look of the business.

Chapter 13
Boats

Boats

Boats, one object that decorators rarely think of painting...

These need painting and a painter and decorator has all the skills needed to do this. It's a massive market and you can make good money.

I have a passion for boats, my dad is a boatbuilder and we have two boats in the family, an old wooden one and a steel barge.

First, let us have a look at what kind of boats are out there and where you find them. Typically, there are commercial and private boats. Commercial boats are things like fishing boats, commercial tankers and container ships. Private boats are sailing boats, power cruisers and narrowboats. Also in this category are Dutch barges.

Above - Fishing boats and pleasure cruisers

Boats

Above - Dutch barges are massive

Above - Narrowboats

If you look at the pictures above, you can see there is a fair amount of painting on a typical boat. On most boats, it is just plain paint. However, on narrowboats, there can be some lines added. These can be done with tape and are easy to add.

Another aspect of the boats that you cannot see in the picture is the bottom of the boat. The hull needs painting regularly depending on the type of paint used previously and the usage of the boat.

Where do you find these boats and how do you get the work?

Commercial fishing boats will be moored in a harbour or port. These will have someone already painting them because it is so important to maintain the hull on these types of boats.

It would be worth approaching the company that currently does the painting and ask if you could subcontract to them. You would learn a great deal about the job and make good contacts with the boat owners.

Private yachts and motor cruisers can be found in marinas up and down the country. I would visit the marina and chat with the people that run it to find out if they already offer a painting service. They might do. If they do, then they will not want you working in their marina competing against them.

If they do not have a painting service, then this is your chance to work with them. They will not let you just set up shop, you will have to strike a deal with them where they get a cut of what you make for using their marina. It will vary from place to place but I would not try to play hardball at first and accept the deal offered (unless it's not viable).

You may need to convince them that you can deliver the goods. To do this you may have to find a couple of boat owners and offer to paint their boat at cost in return for using the photos

Boats

and videos that you take for marketing. You can also get a couple of endorsements for your skills.

Marinas are expensive from a boater's point of view so if they are moored in a marina, they have money and they can afford it. Typically, a marina will charge £200 - £300 per month to moor your boat depending on the size of the boat.

Boatyards are another place that boats can be found. These are more rough and ready and the boaters are more likely to be a poorer crowd. Don't underestimate them, though; at our boatyard, one or two of the scruffy-looking guys are actually millionaires. Same approach as with the marina, chat with the boatyard owner.

When it comes to doing jobs on their boats, people are a jack of all trades. They will attempt most things, they will have a go at wiring, mechanics and joinery. Very few, though, will paint their own boat. This is because it is difficult to get a good finish unless you know what you are doing. This is good for us.

Many boaters are retired and want to spend their time enjoying their boat and do not want to roll their sleeves up and scrape a hull.

How much would you charge?

This is a great question. I have a good idea what the going rates are at the moment but if you are thinking of doing this then you need to do some research as prices will vary from area to area. I am in the north-west of England so not top prices but not backwater either.

Our boatyard has a service that sandblasts and paints the bottom of your narrowboat or barge (they are based on the

Boats

canal) and for this, they charge about £2,000. This usually takes two guys two days so it's good money. They have the overheads of the yard. They get a steady stream of work through the summer. The place where they work is outside but if they built a shed or large polytunnel over the work area then they could handle more work because they would not be affected by the weather.

Painting the bottom of the boat is a good business because it has to be done to keep the hull in good condition. It's not really optional.

Another thing about painting the bottom of the boat is that it is mostly underwater, so you don't have to have as much finesse as a painter when doing this work.

Topsides take a little more care and this is what people see when they look at the boat. Customers will want a great finish. I will talk about paint and application methods later. The cost of painting the topsides is similar to the bottom but can be more depending on the amount of work.

Typically, prices range between £2,000 and £5,000 to paint the top of the boat. This will take you longer to do and you need to be undercover for this kind of work. A floating polytunnel is the best approach where you sail the boat into the tunnel, which is floating on pontoons either side of the boat.

These can easily be made yourself with the permission of the boatyard and this will allow you to work all year round in a nice controlled and reasonably secure environment. If you had electricity you could even heat it.

Boats

Above - Floating painting workshops

What kind of paint and preparation?

Boats are made from either wood, steel or fibreglass. Generally, these surfaces have already been painted so there is no priming process.

First, let's discuss the hull, the bit that is actually in the water.

Boats

The boat is either in fresh water or salt water.

Wooden hulls last well in salt water (it helps preserve the wood) but do like it in fresh water (like a canal). Steel hates salt water because the salt speeds up the rusting process; it is better in fresh (although it still rusts). Fibreglass is fine in both fresh and salty and, for this reason, is the material of choice for many boat owners.

Steel hulls are usually painted with 2-pack paint. This will last the longest in both fresh and salt water. If there is a one-pack paint on the steel, then that will have to come off by sandblasting before 2-pack paint can be applied.

A lot of narrowboat hulls are painted in black bitumen paint because it is cheap and easy to apply. It is flexible too so that helps protect the steel. It needs doing regularly, every year or two. Commercial "for hire" narrowboat companies will bitumen the hull twice a year.

Wooden and fibreglass boats in sea water are painted in antifouling paint. This is available in a limited range of colours, typically black and red and prevents weed from building up on the hull of the boat. It is not expensive, around £80 to £100 per 5 litres.

Above - Hempel are a well-known marine brand of paint

The topsides, the cabin and the hull above the waterline can be painted in much the same way that you paint a house. You could even use standard undercoat and gloss. I wouldn't, because marine paints (even though they are expensive) are great to use, they flow well, and they offer good protection.

There are several marine paints out there, but I will recommend International. This is widely available both online and at marinas. Shop around and get a trade deal, marinas are selling to the public and they are top-notch prices, you want to avoid getting your paint from them as this will eat into your profits.

Above - Toplac is great for the finishing coat on the topsides

Brush, roll or spray?

This is the big question. How do you want to apply the painting on the boat? Believe it or not, I have done all three, but spraying is the least used. The reason for that is at my boatyard there is no indoor space to work and all the boats are in close proximity. This means that overspray could be an issue for me.

Many of the marine paints brush and roll well. I have brushed the cabin on a Dutch barge and the finish was almost as good as a sprayed finish. When painting the hull it's just as easy to roll, the finish is not seen and it's almost as quick as spraying. I don't

Boats

want to put 2-pack through my sprayer, but you could use a simple compressor and cheap spray gun. That's what our boatyard does.

All brand new boats are painted to a spray finish and customers expect that. You just need somewhere that you can do it where it is controlled but you would need that anyway.

I feel like I should do this myself and I have thought about it a lot, I get lots of offers to paint boats. These days I am too busy with the training academy and I have little time for anything else.

If you are interested in boats I have written a book called "Boat Life". It's not about painting, it's about living on a boat and I wrote it just because I know a lot about the subject. Check it out on Amazon.

Boats

Company name: The Boat Painting Company

Website: www.theboatpaintingcompany.co.uk

Tagline: Painting perfection above and below the waterline.

Pitch: We have been painting all types of work including boats for many years and our expertise is second to none. The problem that boat owners have is finding a reliable painter who can finish your boat to the standard that you expect. Our last customer had his boat featured in "Narrowboat magazine" and the pictures of the painting looked amazing. The couple were over the moon with the finish.

Chapter 14
Airless spray plastering

Airless spray plastering

Airless spray plastering is relatively up and coming in this country but not really a new thing elsewhere. I speak to many people (not just decorators) and they have never even heard of it, some even think that I have made it up and it is impossible.

What is it?

Spray plaster is more like a filler than a gypsum plaster. It dries in 24 hours and can easily be sanded to a super smooth finish. However, once it has cured (much like paint) and has been painted then it is rock hard and is very durable.

You can achieve a level 5 finish (which is more or less perfect) more easily and quickly when using the spray plaster process. It is something that decorators can learn, and it means that they control more of the finishing process.

There are several advantages: speed, coverage, flexibility, quick-drying, pre-mixed, no wastage, dries white, no mist coat required, easier to apply and a level 5 finish. Let us look at each advantage in turn.

Speed

Let's face it, applying plaster by hand takes us back to the dark ages, why are we even still doing it? If you can apply the plaster with a sprayer, it will be easier and quicker. For decorators already spraying with airless sprayers, this is a small step to spray plaster. For plasterers, it's a bigger step and they find it harder to embrace the new process.

I have seen both decorators and plasterers do our spray plaster courses and the painters get up to speed quicker. The plasterers, in fairness, are better at smoothing the walls off.

Coverage

One benefit of spray plaster is that it will cover almost any surface, plasterboard, blockwork, Artex and even woodchip. If a customer has an Artex ceiling and they want to get rid of it then it is an easy process to spray plaster over the surface and sand it down to a perfect finish ready for painting.

Flexibility

I have tested spray plaster by applying it to a thin plywood panel. Then allowing it to dry and cure. The board can be flexed, and the finish stays on the plywood. Traditional plaster is a brittle product and cannot withstand any movement.

Quick-drying

When a new build is plastered with traditional plaster then it is not ready for painting for a couple of weeks. It has a high-water content and takes a while to properly dry out. Spray plaster can be applied on day one, second coated on day two and sanded and painted on day three if the drying conditions are good.

Pre-mixed

The plaster comes ready mixed so there is no need to mix it or calculate how much water is needed. It comes in bags that are easy to load into the hopper of the sprayer or in large tubs. Most spray plasterers use the bagged material that works well with the hopper. For smaller jobs, the tubs work fine.

No wastage

The product does not go off as traditional plaster does, it can be returned to the tub and used again the next day if needed.

Sandable

As decorators, we are all used to sanding poorly applied plaster. However, it is not great to sand because it has not been designed for this. Spray plaster sands easily, and if you use a dustless sander like a Mirka Leros it is a quick and easy task. You will get better and better at applying the product and I have seen spray plaster applied so well that it needs no sanding.

Dries white

The product dries white, so it is much easier to cover when you paint it and the area that you are working in is nice and bright too.

No mist coat needed

This is a big bonus to the decorator because it saves us a whole process. Maybe one to keep quiet about! Another bonus is that the paint adheres much better to spray plaster than it does to traditional plaster.

Easier to apply

Plastering is not easy; it is a very physical job and over the years (as you get older) it gets harder and harder to do. The big advantage of airless spray plaster is that the sprayer does all the hard work getting the product onto the walls, you only have to smooth it out.

Level 5 finish

This is one of the strong selling points of the spray plaster: the finish is perfect. Customers are expecting more and more from us and, as decorators, we struggle sometimes to get the plaster

Airless spray plastering

good enough to paint. Here we have a system that is easy to use and gives a great finish.

How much can you earn?

With all these advantages in mind, this makes a perfect specialism for the decorator. You can offer a full package of spray plastering and finishing off with paint. You would become a finisher and the customer would only have to deal with you for the final stages of the job.

Another advantage of this specialism is that for every job you are effectively doubling your business. Typically, plasterers will charge between £8 to £12 per square metre to plaster. That means if you have a house to do and it's 400 metres squared of wall area then that is £3,200 for the plastering (using the lower £8 per metre figure) and for the painting, if you went in at £6 per metre, then you would be getting £2,400.

The big plus is that because you have done the plastering, all the preparation has been done and your painting price is almost pure profit, especially if you spray it.

There are a few objections that decorators come up with against the spray plaster approach, these mainly centre around the cost of the equipment. This is a fair point and to get set up with the sprayer and hopper you could be spending £5,000.

A couple of things to remember when buying the equipment. First, it will be set against your tax and it is a capital expenditure so if you calculate the £5,000 over 5 years then it is only £1,000 per year or £20 per week.

Airless spray plastering

Second, you can always sell the equipment if it is not for you. In my experience, the Graco Mark V (which is the main machine used at the moment) are easy to sell second-hand.

You would have to get up to speed skill-wise and the best way to do this is to do a course. Knauf does product demo days where you can try out the process. However, the best place to go by far is PaintTech Training Academy, who have a City and Guilds assured course in the process.

I know, I would say that, wouldn't I? Well, I couldn't resist.

Check us out on www.painttechtrainingacademy.co.uk

Company name: Spray Plastering Specialists

Website: www.sprayplasterspecialists.co.uk

Tagline: Perfection for your walls

Pitch: Have you ever felt that plastering could be better? Yes, so have we! We have been sanding down poor plastering for years but now we have found a solution. Spray plaster gives a level 5 finish on your walls and ceilings. All our customers are amazed at the finish and also how much quicker the work is completed to a stage where the client can move in.

Section 4

The moneymakers

There are some specialisms that I feel you can make good money in, for various reasons. Some may involve a machine that does the work for you so that you are making money while you sit at your desk.

Some are good money because there are few people doing it and there is a big demand and finally some are good money because they are the next thing and decorators just do not know that they exist.

Chapter 15
Computer-aided signmaking

Computer-aided signmaking

Once upon a time, signwriting used to be part of a decorator's job. It was taught as part of the decorating course (until 1983) and decorators of old used to specialise in it.

Back then, the decorating department at college was in the art department. In Preston, it was in the Harris College of Art. Decorators learned about colour and design and also about lettering.

When a sign was made, it was usually made of wood and then painted, so this was a painter's job and then it was painted using brushes and specialist paint. Then a few things happened all in quite a short space of time.

First, the decorating department was moved into Construction (a bad move in my opinion) and a few things were taken out of the decorating syllabus. One of these was signwriting. It was made into a separate qualification for people who still wanted to do it.

Second, computer-aided signmaking was invented at around about the same time (the early eighties). This meant that you could design a sign on the computer and then send it to a cutter that cut the design out in a plastic film that could then be stuck onto the van.

This system was faster and cheaper and was the beginning of the end of signwriting as a mainstream trade. In the early days, the software was very clunky, and the cutters were expensive. Over the next 10 years, software got easier to use. Windows 3.1 was released in 1992 and suddenly, you didn't have to be a computer genius to operate a computer.

Computer-aided signmaking

I started at college in 1992 and we bought a system to make signs. We were offering a signwriting course and we wanted to also teach the next generation of signmaking using a computer and cutter.

I was lucky I suppose in that when we got the signmaking software (FlexiSIGN) I was trained how to use the computer, the signmaking software and the cutting machine. Suddenly you could cut letters out for a van and stick it on in a fraction of the time (and skill) it took to signwrite a van.

Sign makers would charge similar money to do the work so profits went through the roof. Back then the early adopters of this new technology made a fortune.

What was lost, though, is the signwriters' flair in setting out the sign and some of the unique look of the old signage. New computer-aided signs started to all look the same with little skill applied to the layout.

This technology is still available today, although everyone has forgotten it used to be painters that made signs. It is not that expensive to buy and you can make good money making signs for various businesses.

You would need a computer with some sign making software on it. I will recommend FlexiSIGN Pro because that is what I am familiar with. There may be better ones out there these days.

Computer-aided signmaking

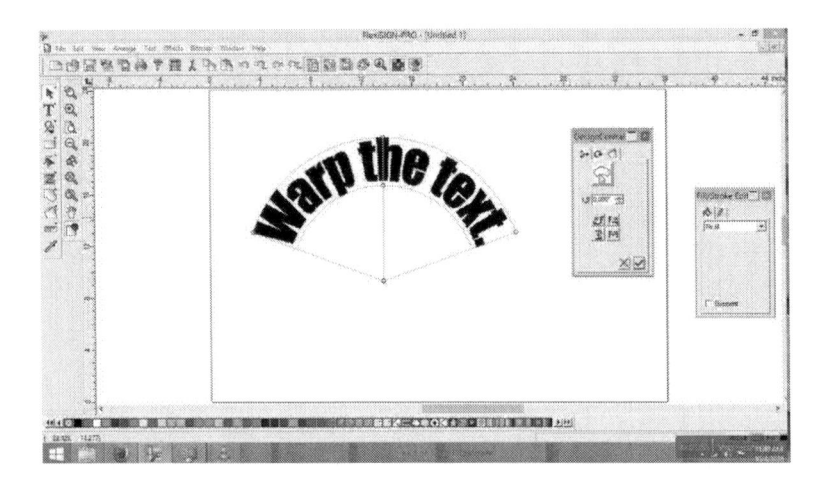

Above - FlexiSIGN software

Above is what the software looks like, it is easy to use. Much like "Word" but with more powerful text manipulation tools. Plus it can turn the design into a file that the cutter or plotter can understand.

The plotter/cutter that we used at college was a Graphtec. These come in various sizes (widths) and will either plot the design onto paper or cut the design out of vinyl.

Computer-aided signmaking

Prices of plotters range from £1,400 for a small one to £6,000 for a large one. I would start small and build up.

The process is straight forward. You design the sign, then send it to the cutter that cuts the design out of vinyl.

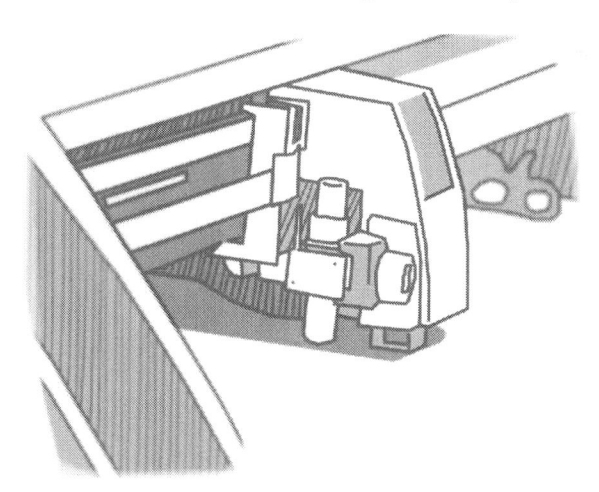

Above - The cutter

Page | **129**

Computer-aided signmaking

Once the design is cut you have to peel away the vinyl that is not needed and leave the lettering on the wax paper backing.

Above – weeding out the waste vinyl

Once you have done this you stick application tape on top of the vinyl. This is like wide masking tape and makes the application of the letters easier. You need to get the letters off the wax backing paper but keep the design intact.

Below you can see the letters being applied to a window. The application tape is being removed now that the letters are stuck to the glass.

Computer-aided signmaking

This could sit well with your decorating business because often if a business is getting their premises decorated, they are also looking at new signage. You could offer this as an additional service but look to expand.

You could just jump in and decide that this will be your specialism and get some equipment and make signs. You would need somewhere to work from, ideally a small unit, but the equipment is small and clean so you could work from your garage or even a spare room at home. You would need a large table or bench to work with the vinyl as it comes off the machine.

As well as signs and "A" boards, you could also offer a service to apply vinyl to vans. Here are a couple of typical vans.

Computer-aided signmaking

Above – a van livery

Above - A simple but effective design

It would probably be worth getting a couple of sign design books to get your head around the layout of lettering. There is not much out there on this, but it is worth subscribing to Signcraft magazine, this is either online or you can get the magazine in print form.

Computer-aided signmaking

This is full of hints and tips on signmaking. There are loads of examples of cool layouts and pricing guides too. This is very much an American magazine because this is big business over there. Well worth checking out if you are serious about this specialism.

Computer-aided signmaking

Name: Superb Signs

Website: www.superbsigns.co.uk

Tagline: Smart signage specialists

Pitch: We are very experienced in signmaking, but we are also decorators. This means that we can paint your facia and then provide the sign. We pride ourselves on having good knowledge of colour and design and we can deliver a sign with more flair than your average computer-generated sign. Many customers come back to us because of our unique design approach.

Chapter 16
Intumescent coatings

Intumescent coatings

When a large building is built out of steel the steelwork needs to be protected against fire. If a steel-framed building sets on fire, the steel will twist and warp when it gets hot.

This causes all the doors to jam up and no-one can get out of the building. For this reason, the steel needs to be painted with fireproof paint. These are called intumescent coatings.

The way the coating works is that when it gets hot it foams up and keeps the heat off the steel for some time, this time gives people a chance to get out of the building before it twists and warps and falls apart.

Generally, the intumescent coatings are applied by airless sprayer. You need a decent size sprayer to do this job because the product is thick. These days products are getting better and better and some products will spray through a smaller machine.

Before I go on, I would like to say that I am by no means an expert in this area of work and I have never done it commercially. However, I know people who have, and it pays well. The work is straightforward, and you do not get many of the "snagging" issues that you get with standard decorating.

Typically, the intumescent painter will do the work early on in the build when all the steelwork is exposed, and the floors are in. This means that often you are outside but sheltered. There are no obstacles in your way and there are few other trades either.

Intumescent coatings

Above - Exposed steelwork, waiting for the floors to be fitted

There is a technical element to the job, and you must measure the thickness of the coating using a micron gauge. This is simple to do, it can be measured wet with a comb.

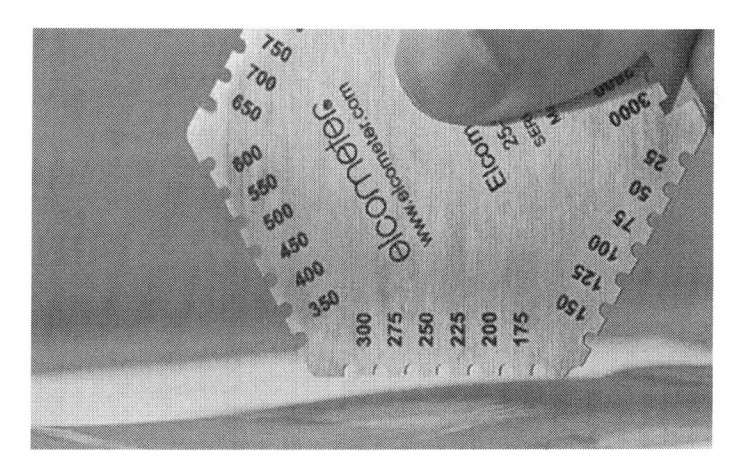

Above – a wet film thickness gauge

Intumescent coatings

It can also be measured once it is dry with an electronic device. It is very important the coating has the correct thickness and different parts of the building will have a different specification so you would have to follow a drawing.

The work must be certificated for the building to pass building control so it important that at this stage you get it right.

There are several products on the market and the job you are pricing for will have likely already have a product specification. However, if you specialise in this area there will be an expectation that you have good product knowledge.

All the product manufacturers have training programmes to teach you how to use their product and to teach you coverage rates, application methods and typical pricing rates.

Above – an example of an intumescent paint

Finally, if you specialise in this kind of work then you must invest in a sprayer. Although in theory you could brush or even roll

Intumescent coatings

intumescent paint, you would not want to. An airless sprayer will be faster and give you better results. You will need a decent-sized sprayer, and the market leader at the moment is Graco.

Above – the very nice looking Graco 695

The Graco 695 is a great sprayer for spraying the intumescent paints. You could go for the Graco Mark V, which is very similar but designed to pump thicker materials such as spray plaster. However, it would be more expensive.

Intumescent coatings

Company name: Intumescent Coating Specialists

Website: www. Intumescentcoatingspecialists.co.uk

Tagline: We are reliable and deliver to programme every time.

Pitch: We have been working with fireproof coatings for many years and we pride ourselves in delivering on time, every time and to specification. Our last client was developing a £60 million building that had to be completed by a tight schedule and we completed our element of the work a week early and to specification and with no comeback later in the build. The client was over the moon.

Chapter 17
Murals – painted, multi-plate and one-piece

Murals – painted, multi-plate and one-piece

Bespoke is where it is at these days, this is what people want. A unique wall in their house that no-one else has. Something personal too, maybe a picture of a romantic haunt, a beach scene where you had your first holiday or an image of a special event such as a wedding.

Murals fulfil this need for a special wall. The idea has been around for many years.

Above - A Stone Age mural

Above - A more well-known ceiling mural

Murals – painted, multi-plate and one-piece

There are three ways to produce a mural and each way is less time consuming than the last.

The original way to produce a mural is to paint it on the wall. This method is still used, and companies will pay over the odds for even the most mediocre hand-painted mural. Hand-painted work is the ultimate bespoke image. There is literally only one. Even if it is recreated at another property it will never be exactly the same.

Above - A hand-painted mural

It need not be that difficult, you can keep it simple, choose a design, draw a grid on the design and then draw a bigger grid on the wall. Use the gridlines as a guide to enlarge the image that you have chosen.

You can still use emulsions to do the job. But if we are honest, if we are not artistically inclined then we won't go down this road.

Murals – painted, multi-plate and one-piece

The next way to produce a mural is to buy one as a multi-plate mural. This comes in several squares that have to be matched on the wall.

I found one on Amazon and here are the details,

- Covers any area up to 8ft x 10ft
- Wall Mural comes in 12 individual panels
- Comes rolled up in a similar way to wallpaper
- Interactive wallpaper murals made especially for children
- Free Wallpaper Paste Included

First, 8 foot by 10 foot is not very big. A typical lounge wall will be 12 to 14 foot.

Second, let's face it, 12 panels! Nightmare. How hard is that going to be to put up? What are the chances of a step in the image? Pretty high, I would guess.

If you ruin one panel for whatever reason, then the whole mural is ruined.

Multi-plate murals are the way many companies have gone, and decorators will attempt to hang these without question. However, there is a third way, and this is the ultimate.

One-piece murals

This is printed on one piece of wide vinyl. Any size (more or less) and any image. Even a bespoke image. There is a company very local to me (he shares our building) called Vista Digital and they print the murals for you. They have access to thousands of images, and you can even send your own high-resolution image to be used.

Murals – painted, multi-plate and one-piece

Above - An image being printed

The material that is printed onto has these qualities:

Fireproof
Water-resistant
Scratch-resistant
Tough and durable
Easy to clean

This makes it ideal for a commercial environment such as a hotel reception area.

Below are a few examples of murals installed in different commercial properties.

Murals – painted, multi-plate and one-piece

Murals – painted, multi-plate and one-piece

Above - This one is on the stairwell at Sotheby's in London

There are a few advantages to this specialism for a decorator and they are:

First, they are easy to install. Because they are one-piece, usually you can hang the mural on your own. The product is a wide vinyl, so you paste the wall, and also it is very tough, so you do not have to be too careful when hanging.

Second, it pays well. Typically, a wall is 4 metres by 3 metres (12m^2) then the going rate is around £15 per metre. That is £180 for the install. You could comfortably install a mural that size in a morning.

Third, it's a unique product and will set you apart from other decorators if you offer it as a business.

Murals – painted, multi-plate and one-piece

Company name: Bespoke Mural Installers

Website: www.bespokemuralinstaller.co.uk

Tagline: Your very own unique wall

Pitch: We have been installing murals and wallpaper for many years and we have found that traditional murals are very poor quality, and it is very difficult to avoid problems with the joints or seams. They either lift or there is a step in the image. Our one-piece mural avoids this and gives a perfect wall every time. One customer stated that it was "simply unbelievable."

Chapter 18
Architectural films

Architectural films

What are they?

Blue Peter was famous for making things on the show and one of the "go-to" materials was "sticky back plastic". Fast forward to the 21st century and this stuff has got sophisticated. These days it is very hardwearing and looks and feels like the material it represents.

Architectural films are vinyl-based textures and finishes that can cover most surfaces. You can apply the films to all the surfaces that you currently paint, for example, windows, bar fronts, doors, lifts and even within a bathroom or toilet.

There is a massive range of colours and designs.

From the customer's point of view, there are a few massive advantages. Let's say that you own a hotel, and the doors are prefinished, and they are looking a little dated. You have several options.

You can paint them to freshen them up, but if you did this you would have to paint the whole room. Plus, many hotels prefer prefinished doors because they stand up to wear and tear better than paint. Painting the doors would also be very disruptive and the area would be out of action while the painter painted the doors and then allowed them to dry.

Another option is to replace the doors with new ones. This is an expensive option. You have to buy new doors and pay a joiner to fit them, you also have to skip the old doors which these days is not environmentally friendly. The process again is disruptive, you would have to take the area out of commission while the joiner removed and replaced the doors.

Architectural films

With architectural film, you would keep the existing door and wrap it in the latest fashion, let's say a nice rustic oak. This process would take minutes instead of hours and could be done after the resident checks out and before the next one checks in. Zero down time for the hotel. A whole floor could be refurbished in an afternoon.

Finally, from the hotel's point of view this is a very cost-effective solution. Compared to the cost of a new door, the cost to hang it and dispose of the old one and also the cost of not being able to use the room for a customer, the architectural film can cost seven times less for the customer.

From the installer's point of view' this is a very lucrative business because the installation is fairly straightforward and fast, no special tools are needed and for a decorator used to using wallpaper and wide vinyl, it is a familiar process.

The replica finishes are almost impossible to distinguish from the real thing. So, an oak architectural film looks like oak, it feels like oak too when you touch it, there is a texture and a grain, and it feels like wood. You must see and feel it to believe it.

The following are advantages of architectural film:

Versatile

It can be used on a wide range of surfaces including chairs and tables, metal and powder-coated surfaces, skirtings and architraves. When refurbishing a room, the dressing table and wardrobe can be made to match exactly the architraves and windowsills.

Realistic

This is the most astonishing thing about the product. I worked on the refurbishment of a restaurant. We were doing two floors and there were a lot of tables. The owner had the tables wrapped in architectural film. The tables were shabby and I was sceptical about the process. The tables were wrapped in a very rustic wood finish and once completed they looked like brand new high-end tables. This would have cost an absolute fortune had he gone out and bought new tables.

I knew that the tables were wrapped in vinyl but to the untrained eye, they looked like wood laminate tables. Not just visually either, they felt like real wood.

Cost-effective

As previously discussed, the cost of using architectural film over replacing can be up to 4 to 7 times cheaper depending on the actual situation. Costs saved in the lack of downtime alone can be worth thousands. For example, the restaurant did not need to close and the tables could be wrapped during normal closure time.

Eco friendly

These days this is big, especially from a marketing perspective. More and more people do not like to create waste or throw things away. That you are refurbishing the item and not having to waste it makes it a popular choice with customers. This is only getting bigger as more people jump on the environmentally friendly bandwagon. This makes this business much more future proof.

Durable

One disadvantage of both paint and wallpaper is that it is not very durable. People are going for the durable matt emulsions so that they get longer out of the decoration. Architectural film is highly resistant to abrasion, scratches, and impact. It is also resistant to different temperatures and humidity (bathrooms and kitchens for example). Even in high traffic areas, the product will last years. This is a massive advantage to the customer.

You can apply it to any surface

You can apply the material to wood, aluminium, painted surfaces, powder-coated surfaces, laminates, limestone, plastic, stainless steel. This makes it even better than the legendary multi-surface paint that needs no primer.

A massive range of designs

You are spoilt for choice with what you can offer your customers. You can do plain colours so, for example, you could wrap uPVC windows in an anthracite vinyl wrap, it looks like it has been sprayed and you have no masking!

There is a massive range of metallic finishes to rival any car respray, this could be applied to doors or reception desks. A massive range of wood finishes which look like the real thing and marble and other faux finishes such as copper and granite.

Easy to apply

For us, this is the icing on the cake. Not only is it cost-effective for the customer and looks amazing, but it is also easy to do. With a few tricks of the trade, you can be applying the architectural films on large flat areas without bubbles and

Architectural films

around curved surfaces with ease. This could be the future for a lot of refurbishment projects and you can offer it as a specialism then you could be quid's in.

Company name: Dramatic Transformation Decorators

Website: www. dramatictransformationdecorators.co.uk

Tagline: Any finish on any surface

Pitch: Paint and wallpaper can only do so much. If you want a complete transformation then we can wrap most elements of your property to look like brand new, high-end substrates like oak and marble. Our customers not only save a fortune, because they need not renew everything, but they also have minimum disruption as the refurbishment process is quick and painless.

Section 5

A bit arty

A lot of decorators are a bit arty and one reason they became decorators is because they were good at art and they enjoyed painting.

This section is not about the money, it is about doing something that you enjoy if you have an artistic streak. I was a signwriter at my original firm and I suppose that makes me a little arty.

I feel we have let go of the artistic side of the trade and become "construction" and this is a shame.

Here are some artistic specialisms.

Chapter 19
Decorative finishes

Decorative finishes

If you have been decorating for any length of time or if you have attended college to do a decorating course, then you will know what decorative finishes are.

They are things like stencilling, wood graining and marbling. But also rag rolling and sponge stippling. Now, I know what you are thinking, "Oh my god, no-one has that done these days" and "why did we ever learn that at college, it was a waste of time".

In some ways, I could agree. However, these techniques have always been a part of our trade and we should be proud of the skills needed to carry them out.

In the 90s it was all the rage, it was on the telly with "Changing Rooms" and I had hundreds of people coming on our decorative technique course for many years. Everyone was having wood-grained doors and marble tables. Rag rolled and sponged walls too.

It fell out of fashion as everything does, but that was 20 years ago and these things come back in cycles and I would not be surprised if this kind of work came back but in a slightly different form.

On many of the large building projects, it is now cool to have an industrial, rustic look. The last job that I worked on had fake copper panels in the reception area. These were actually sprayed and cost a lot of many to produce. They looked great!

In the States, they have a more developed decorative finishes market, and many commercial buildings have marble and wood that is painted rather than real. A visit to Disney World in Florida is a great example.

Decorative finishes

If you have ever been it's amazing to realise that everything that you look at is actually fake and has been painted to look real. They have a massive workshop (the size of an aircraft hangar) where all the work is carried out. Full-size trees are made of fibreglass and painted to look like wood.

Closer to home, one of the best places to see how amazing decorative painting can be is Chatsworth house. If you are a decorator, it is well worth a visit.

One feature is a wood-grained door with the painting of a violin on it. This looks so real that it takes some convincing that it is painted. You cannot get close to it as it is cordoned off.

Above – the violin at Chatsworth house

Decorative finishes

Above - A closer look

Decorative finishes

Now I am not suggesting that you would want to do something like this, but it is interesting to see what level of work can be achieved.

One thing that people always ask when discussing painted techniques such as graining is "Why bother, just buy a hardwood door and varnish it". This is a fair comment as these days ordinary people can afford hardwood, unlike in days gone by where the working class had to imitate it.

Sometimes it is impossible to use the real thing, so for example if you have a steel front door but you don't want it to look like steel then you could have it grained.

Another great example are the columns in the Trafford centre.

Above – marbled columns in the Trafford Centre

Decorative finishes

The reason that these were painted instead of using real marble was twofold. First, for columns so large it would have cost a fortune to use real marble. Second, the real marble would have been very difficult to work with. It's much easier to have plaster columns and then paint them to look like marble.

Company name: Decorative Paint Specialists

Website: www. decorativepaintspecialists.co.uk

Tagline: When real is just not good enough

Pitch: We have been producing decorative painted work for many years and we find that our customers love the finish that can be achieved on dull and uninspiring surfaces. The money that can be saved by creating the illusion of an expensive material is what attracts our customers. One shop owner said that not one of his customers suspected that the marble in his shop was not real.

Chapter 20
Colour and design

Colour and design

In the States, an interior decorator does the colour scheme for the customer and even specifies furniture and fabrics. In this country, this job is done by an interior designer. Neither roles are carried out by a decorator. However, I think that if you are arty and creative then this is an area where you could excel.

There are two approaches that you could take here.

First, you could specialise in decorating interiors but make yourself different by giving colour and design advice, maybe even produce mood boards for the customer.

Second, you could just do colour and design work for other decorators and builders so that they can add the service to what they offer.

I would do the first and if I got good at it then I may develop the business to move towards the second option.

When it comes to colour, we are very weak on our knowledge and it is something that a lot of decorators shy away from. Even if you don't make colour and design the area that you specialise in, I think it is still important to understand colour.

I will explore three areas of colour that I think are useful to know. These are: how light affects colour, colour systems such as the BS4800 system and how colour schemes can be put together.

Colour and light

One thing I used to ask my students is: "Where does colour come from? Why is that door red?" A common response would be to frown, look at me like I am daft and say: "It's painted red."

Colour and design

I suppose this is true, but it is not where colour comes from. Because we live in a world of colour it is something that we never think about but there are a couple of things going on when we see colour.

First, the colour that we see is contained in the light that comes from the sun (usually) and the sun's spectrum contains all the colours of the rainbow. These are red, orange, yellow, green, blue, indigo and violet. There are a million subtle colours between these. Yellowy orange and bluey greens.

You cannot see a colour not contained in this spectrum of light.

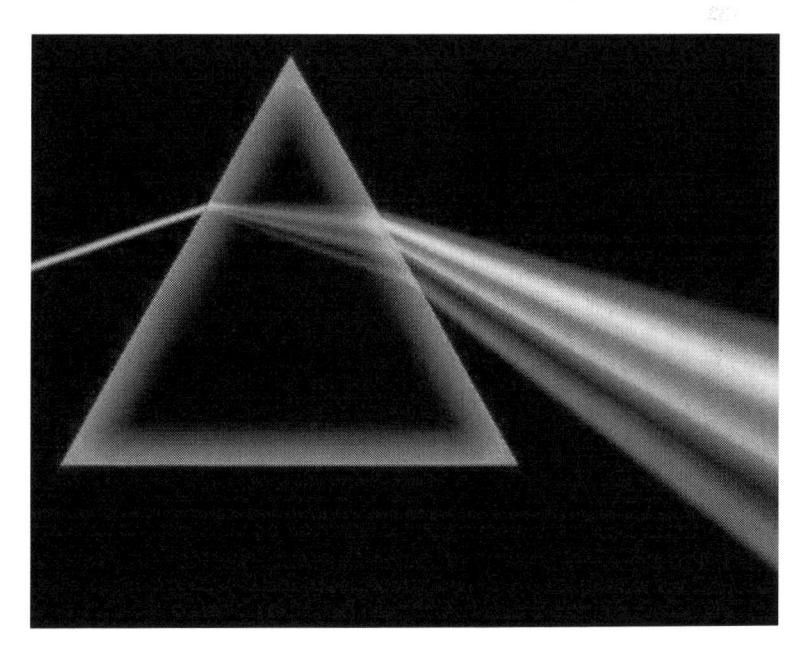

Above – this looks much better in colour

To confuse matters further, when all the colours in light are mixed you get white light. So, on your television, which uses light to create colours, this is what happens.

Colour and design

With paint, however, colour works in a different way. If you mix all the colours of the rainbow using paint, you will definitely not get white. Brown most likely. Interestingly, you cannot get a brown light because of the way light works.

Back to the main plot.

What happens when you look at a red apple is that light (containing all colours) falls onto the apple and the red element is reflected into your eye. All the other colours are absorbed by the surface.

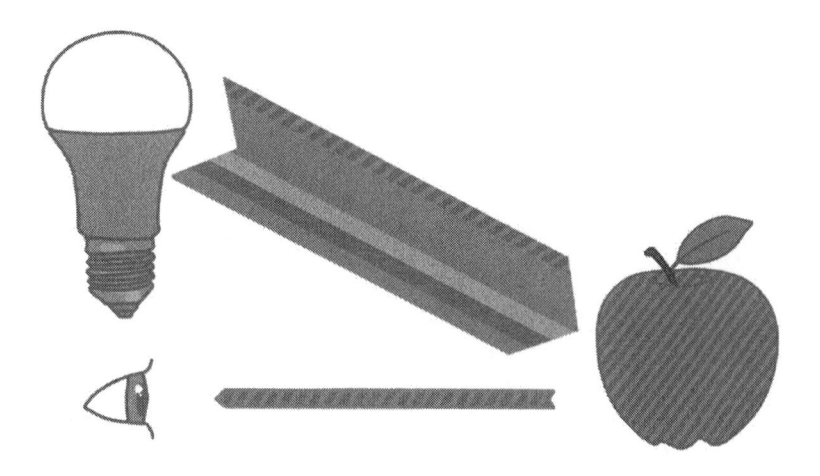

Above – only red is reflected to the eye

Easy so far. Yellow surfaces reflect yellow, blue surfaces reflect blue and so on.

Colour and design

White surfaces reflect all the spectrum (and so appear white) and black surfaces absorb all the spectrum (and so appear black).

Black and white are not colours as such.

Just out of interest there is another piece to the puzzle. Our eyes cannot see the full spectrum of sunlight, we can only see part of it. The extreme edges are invisible to us. So extreme red (infrared) and extreme violet (ultraviolet) we cannot see.

To show you this, if you do not believe me, take your TV remote control and look at the end of it. It should look something like this.

Above – The infrared light on a remote control

Colour and design

Press the buttons while you look at it and although you know there is an infrared light coming out of the end of the remote control, you cannot see it.

Now, get your phone and turn the camera on. Look at the end of the remote through the camera on the phone.

Press any button.

The phone can see the light blinking and therefore so can you. This is because the phone can see a slightly wider spectrum than your eye. It's quite an interesting little experiment you should go and try it now.

I will wait, the remote may be down the side of the sofa. "Who had it last?" would be my question to the kids, if you have any.

So, you cannot see some of the spectrum, which is amazing because you think that you are seeing everything that there is.

There is another aspect of colour that we need to know about as decorators and that is the effect of artificial light on colour.

If you see a fluorescent light in an office, then it will have a different spectrum to that of sunlight. It has the same colours just in different amounts.

On the next page, you can see the spectrum of sunlight, it is high in all the colours and just tails off either end with the infrared and ultraviolet.

Colour and design

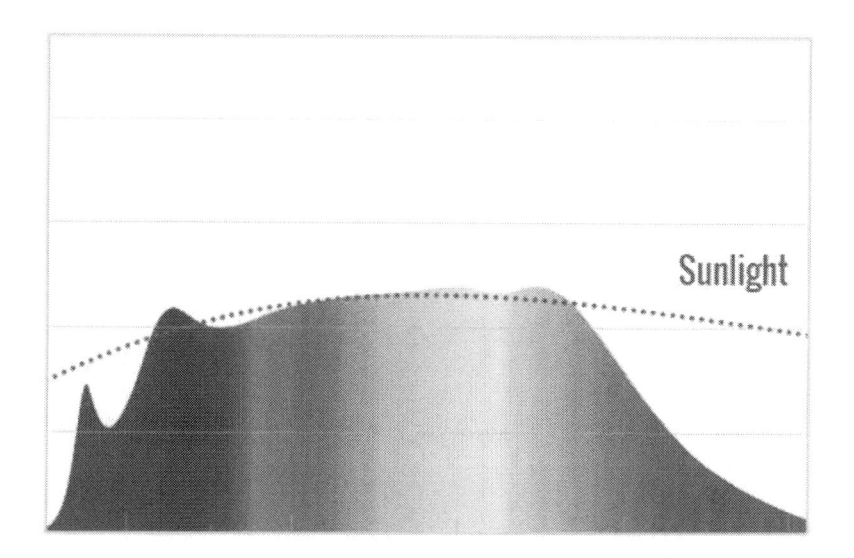

Above – The spectrum of sunlight

Artificial lights will have a different spectrum so, for example, fluorescent lamps typical in an office are stronger on the blue end of the spectrum and weaker at the red end.

This means that under a fluorescent light blues appear brighter and reds duller.

A tungsten filament bulb (which are getting rarer these days) are strong in the red end of the spectrum and weak in the blues.

If you photograph your lounge in daylight and then again at night with the lights on and then compare the two pictures, you will see a big difference in how the colour appears.

Try it; you will be amazed.

It is important to understand that colours appear different under different lights when choosing colours for a scheme.

Colour and design

A final mind-blowing fact about all of this is when you look at a sodium lamp. These are the lights that you used to see on the motorway; yellow lights that are difficult to see under.

Below is the spectrum of a sodium lamp.

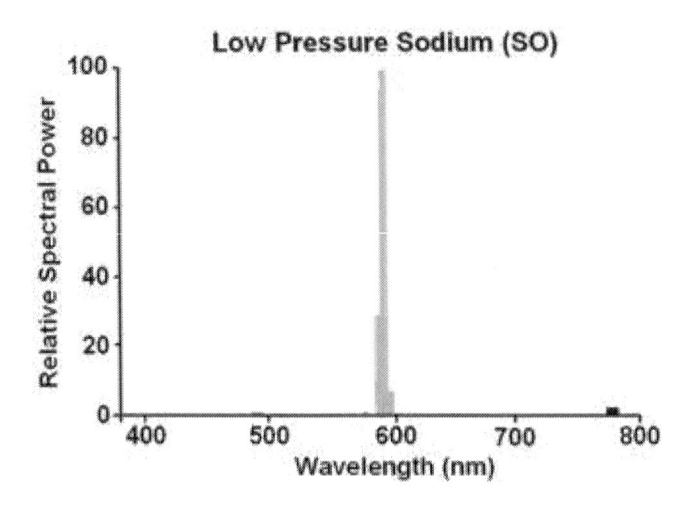

Low Pressure Sodium (SO)

As you can see it only contains yellow, no other colours. This is where we get to test our theory about colour and light.

Why do the following colours appear as they do?

Red – reflects the red element of the light and absorbs all other.

White – reflects all elements of the light

Black – absorbs all elements of the light

Yellow – reflects yellow, absorbs the rest.

Now imagine that you are in a room with only a sodium lamp to see. In front of you are four cards, each one in the colours above.

Colour and design

Card 1 is red but under a sodium lamp, it appears black because it has absorbed all of the yellow from the sodium lamp and reflected nothing.

Card 2 is white; this appears yellow because it has reflected all the colours in the light but the only colour in the light is yellow so that is what we see.

Card 3 is black and appears black. All light is absorbed.

Card 4 is yellow and appears yellow. Yellow is reflected from the surface.

In fairness, you would probably have to see it for yourself, but it just goes to show how lighting can affect colour. These days, thankfully, many lights in our home and workplace are "daylight", so they don't alter colours much.

Colour and design

The BS4800 system

This is a British standard colour system used when specifying colours for buildings. In the old days, when colours were mixed in the factory, it was not practical to have too many colours made. A limited range of colours was mixed and put on the shelves at our local supplier.

Another problem was that Crown might mix "magnolia" different to Dulux. So, a standard was developed that all companies worked to.

Magnolia was 08 B 15, and this was put on the label. In theory, everyone made the same colour 08 B 15.

Many decorators and specifiers got to know the common colours used and just asked for the number. What does the number mean?

Well, the number tells us a lot about how colour works with paint and how colours are mixed using different pigments.

The number is in three parts. Using magnolia as an example (or maybe I should use grey these days) then the number means:

08 is the colour; in this case, yellow/red

B is the greyness; how much grey is in the colour.

15 is the weight and it is how light or dark the grey is that has been added above.

The colour scale is:

00 - Grey

Colour and design

02 - Red/Violet

04 - Red

06 - Yellow/Red

08 - Yellow/Red

10 - Yellow

12 - Green/Yellow

14 - Green

16 - Blue/Green

18 - Blue

20 - Purple/Blue

22 - Purple

24 - Violet

Yes, I know there are two yellow/reds, but this is because yellow/red is basically cream, and this colour is used a lot in buildings.

Colour and design

The letter is a scale from A to E:

A – Grey

B – A lot of grey

C – Some grey

D – A bit of grey

E – Pure colour

So, if you have a colour 04 E 53 then it is pure red. "E" tells you it has no grey at all in it.

00 A 01 is a grey. You know this because it is "00" and also it is "A".

10 A 03 is also grey. We know this because it is an "A"

Magnolia is an orange, believe it or not, a yellow/red with a lot of pale grey in it. If you get an orange yourself and put pale grey in it, you will get magnolia.

Finally, the last number tells us the weight, or how light or dark the colour is. The numbers are paired with the letter:

A - 1 to 13

B - 15- 29

C - 31- 40

D - 41 – 44

E - 49 - 55

Colour and design

To understand this, you need to get yourself a colour card and find these colours.

08 B 15

08 B 17

08 B 21

08 B 25

08 B 29

All five of these colours are "08 B" so that means that they are a yellow/red (08) with a lot of grey (B) in.

The last number tells us how light or dark that grey is. The numbers 15 to 29 are only used with the letter B.

15 means the colour is light (Magnolia) and a very light grey was added to make it.

29 means it's very dark (Vandyke brown) and means that a very dark grey was added to make it. You can see from your colour card all the other colours are somewhere between.

When you get to "C" the whole process starts again. 31 being light and 40 being dark.

This bit is the hardest to get your head around.

These days there are many more colour systems out there, for example, RAL. RAL is used in many industries, which makes it useful. However, BS4800 is just for paint and for that reason I think we should understand it.

Colour and design

My old boss used to memorise the BS4800 numbers for a particular job so that when he walked round the job with the main contractor he would just spout off a load of codes, which made him look very knowledgeable (well, in his mind it did, anyway).

Colour and design

Basic colour schemes

Now that you have some knowledge of colour and you can spout off a few codes to impress Mrs Jones, "Oh yes, I think we will have 08 B 15 on the walls" then you are well on your way to being a colour and design expert. All you need are some colour schemes up your sleeve that you know will work.

I have three that I use.

1. Monochrome

A colour will always go with itself. So, if you use lighter and darker versions of one colour then that will always look good. An example of that was the series of colours that we looked at when discussing BS4800, let's look at them again.

08 B 15

08 B 21

08 B 29

You could have a white ceiling (white and black go with anything), and 08 B 15 walls, 08 B 29 on the woodwork and 08 B 21 on the doors.

Okay, a bit old fashioned perhaps.

You can take any colour and use lighter and darker versions of it, and it will give you a great colour scheme.

2. Complementary

The above approach always works but is a bit boring. To liven the scheme up you can introduce a "complementary colour",

Colour and design

this is a colour that is opposite on the colour wheel. We have not looked at the colour wheel so here is an example of one.

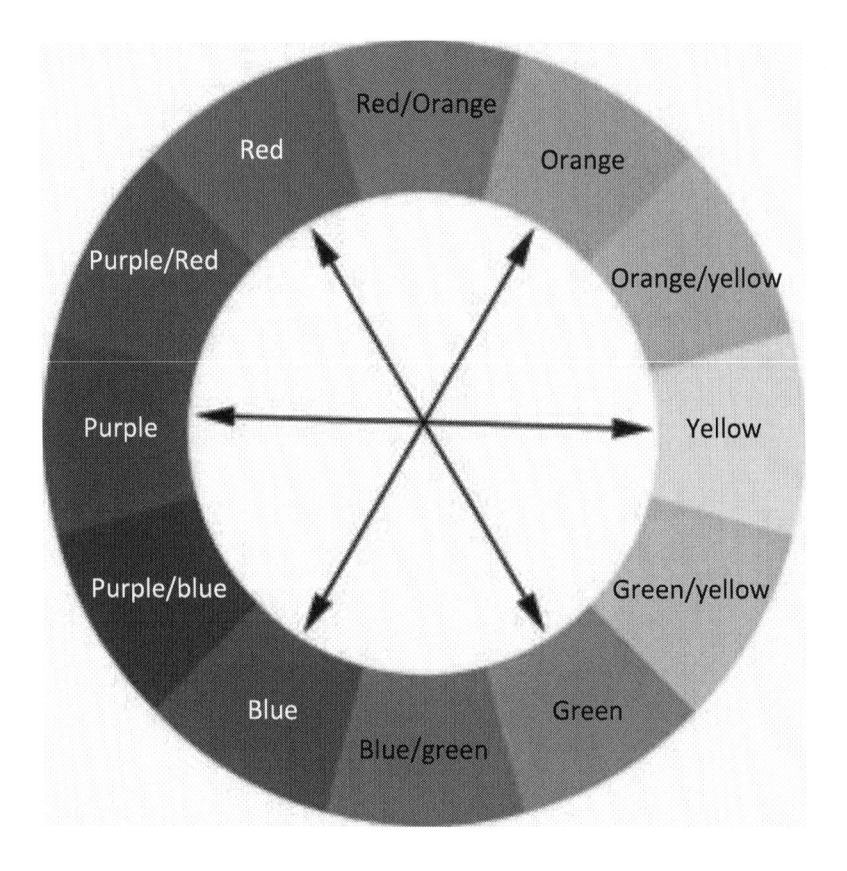

Above - A simple colour wheel showing complementary colours.

A colour wheel is a great way to understand colour. Half the colour wheel are cool colours, and the other half are warm colours.

Complementary colours are opposite on the colour wheel. For example, red and green, orange and blue, yellow and purple.

Now I know what you are thinking – ugh. Red and green? Really? That's not a nice colour scheme.

Well, there are a few things to remember when looking at using complementary colours in a scheme.

First, you rarely use pure colours in a colour scheme, you are usually using greyed down versions of the colour. For example, you would not use bright orange on the walls, but you may use cream, which is basically a dulled down orange.

Second, you would not use the two colours in the same quantity. Complementary colours intensify each other. So, if you put a block of orange next to a block of blue, they would appear to intensify each other and even vibrate.

But in different quantities, for example a lot of orange and a little blue (a blue lamp shade), the orange will intensify the blue (a lot) and the blue will intensify the orange (a bit) and this makes the scheme interesting to the eye.

To sum up, if you did a blue scheme with lighter and darker versions of blue with a splash of orange then that would be a nice, interesting colour scheme.

3. An Analogous scheme

Our final colour scheme is just using colours next to each other on the colour wheel. For example, yellow and orange.

Don't forget we are not using pure colour here but greyed off versions of the colour.

Colour and design

Company name: Decorators with Design

Website: www. decoratorswithdesign.co.uk

Tagline: We add a little colour scheming to your decor

Pitch: We have been decorating for many years and what we have found is that homeowners struggle to choose colours that have the "wow" factor. Most decorators will not help you with this and an interior designer can be expensive. We bridge the gap by giving you solid colour advice on your decorating so that the finished product is the envy of your friends. One of our customers had their decorating featured in Home magazine.

Chapter 21
Window splashes

Window splashes

This is rarely seen in the UK but is big in the States. If you have a special promotion, then you can do a massive banner painted onto the window of your shop front. It is only temporary and can be cleaned off once the promotion is over.

Here are a few examples.

Above – a typical window splash

Window splashes

Above – more window splashes

Window splashes

As you can see, this is a signwriting job. However, the lettering is large and therefore easy to paint plus you can do a drawing and attach it to the other side of the glass and use it as a guide.

In fairness, the guys that do these in the States do them freehand and they do them fast. Have a look on YouTube and you will see one being done.

Because this kind of work comes out of the marketing budget of the company it generally pays well. The same money spent on a radio advert or some brochures will be spent on this.

Another great thing about this (besides the fact that you would be one of the first to do this in the UK) is that it is a recurring income. Generally, if the campaign succeeds then the company will come back for more.

The paints used are water-based and easy to use and easy to clean off. Most window splash artists recommend a window cleaner to remove the artwork once it is finished with.

Company name: Window Splash UK

Website: www.windowsplashuk.co.uk

Tagline: Advertising with a difference

Pitch: Window splashes are big in the States because of their effectiveness in driving a promotion. They are highly visible and unusual. We use washable paints so that your shopfront can be returned to its previous state once the promotion is over. Our customers find they get more results than radio advertising at a fraction of the cost.

Chapter 22
Pinstriping

Pinstriping

This is in the arty section because it takes a high level of skill to get good at it. I have classed it as a decorator skill because that's what it used to be. Decorators used to learn the skill of signwriting as part of their apprenticeship and part of that was painting lines or pinstriping.

It would be worth typing "pinstriping" into YouTube and watch some videos to get a glimpse as to how cool this skill is.

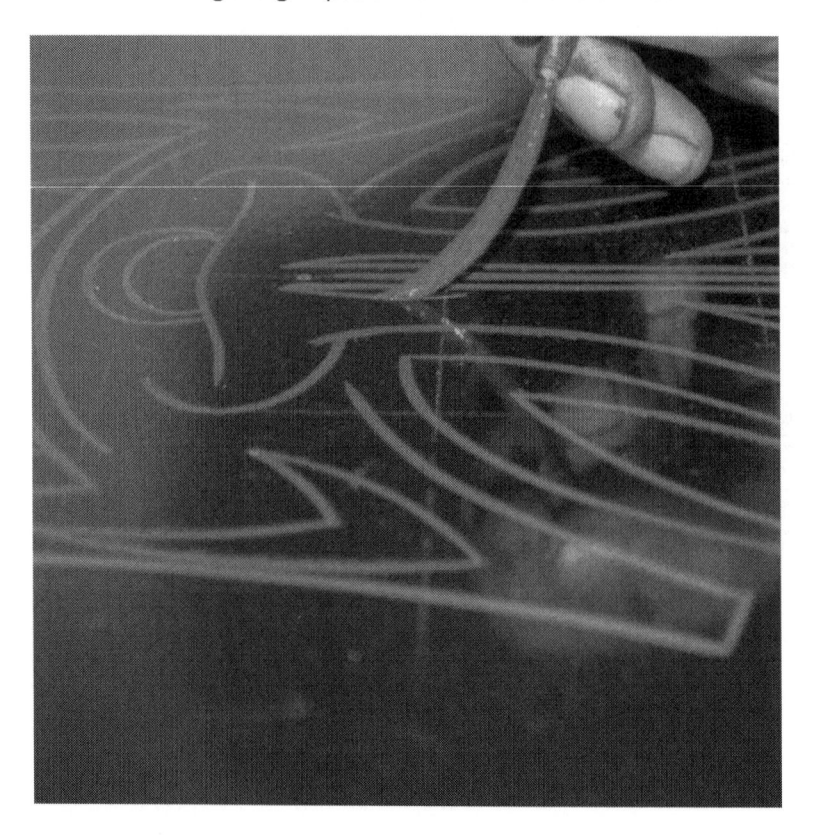

Above – pinstriping is really cool

Pinstriping

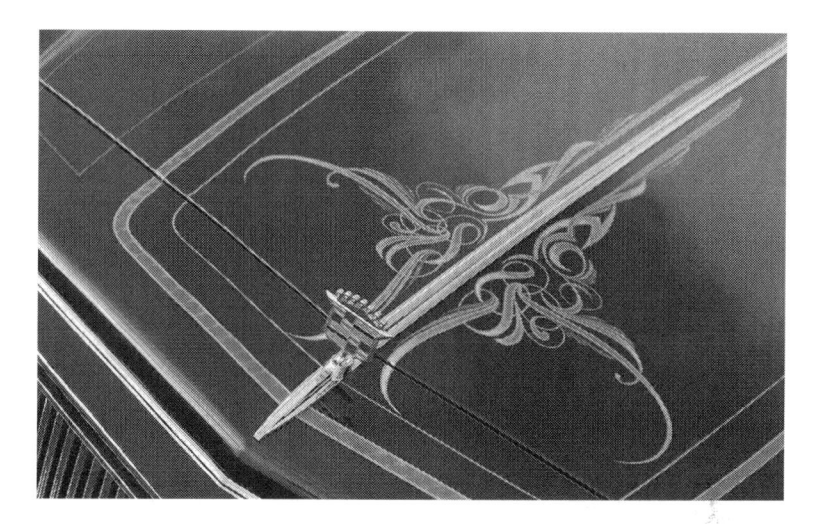

Above - These designs are painted using a pinstriping brush or a sword liner (Below).

Pinstriping

They say that if you pull 1,000 lines with a sword liner then you will be skilled at it. Sounds easy, eh? It is fairly easy to pull a line with the sword liner, as it is designed to make the process easier.

Pinstriping is big in the States where vehicles are decorated with various symmetrical designs. Motorcycle tanks, car bonnets, trucks are all treated to the pinstriping process. The guys that do it are well paid and the fast ones who are well known for their work can earn up to $1,000 per day.

It is something that you would have to build a reputation for, and you would have to build the customer base too. The UK market is different to the American market. However, I think if you were good at it, and you developed some of your own designs, then you could build a following.

These days with social media, pinstriping is a visual art that would be liked and shared. It would also make great videos.

I think it would take a while to build your skill and build your market. However, once you had built a following, I think it would be very difficult for a competitor to move in onto your business.

Another aspect of this, and I will discuss it with the other arty specialisms, is that in these days of computer images and machine prints the hand-done designs will become more sought after as they have a certain quality that is hard to replicate by machine.

This skill is transferable to other arty areas too, signwriting and some of the decorative techniques use the same skills and paints and brushes as pinstriping.

Pinstriping

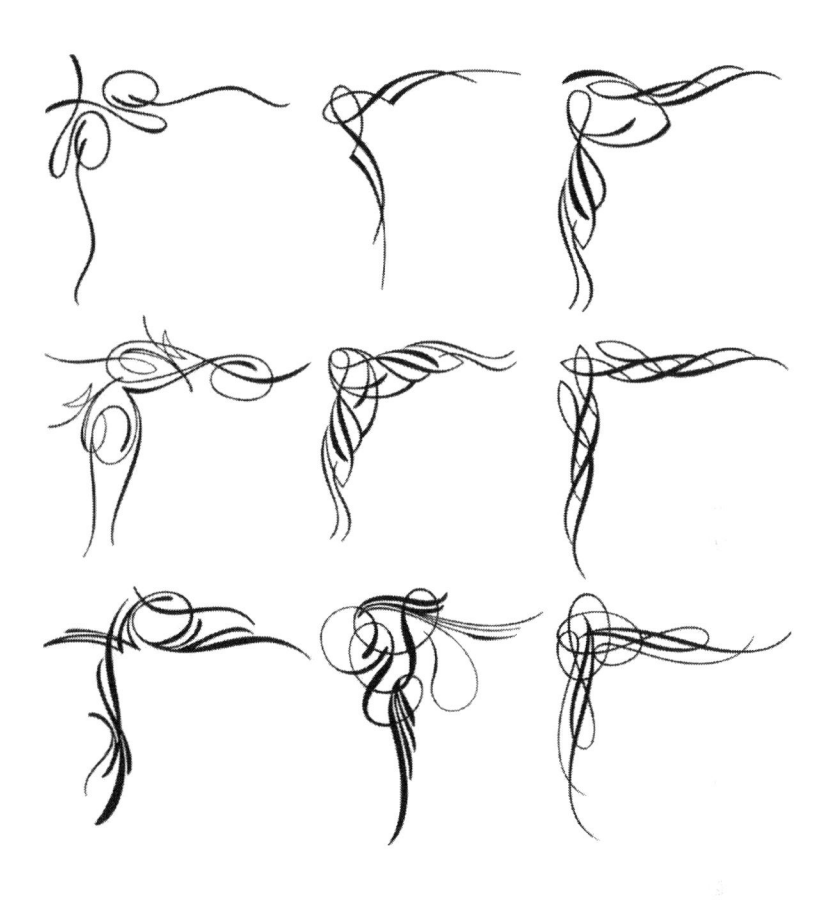

Above – some cool pinstripe designs

Pinstriping

Company name: Pinstripe Perfection

Website: www.pinstripeperfection.co.uk

Tagline: Unique pinstripe designs for your vehicle

Pitch: We have found that for owners of custom or specialist vehicles, pinstriping offers a level of customisation that is hard to match. Our customers like our designs and style, which bring a unique look to your vehicle.

Chapter 23
Gilding

Gilding

This is the application of metal leaf onto a surface for decoration. We typically think of gold leaf when we say gilding, but it can be any metal such as silver or bronze.

If you want something to be gold, for example, the tips on railings, then you have a couple of choices. You can use gold paint, or you can use gold leaf. Gold paint will not last and will go dull and lose its lustre quickly. Gold leaf, however, will last a long time.

Above - gilded tips on railings

The drawback of gold leaf is the cost, it is more expensive than gold paint. However, the look of it outweighs the cost.

Gilding

Gold leaf can be bought in little books containing 25 leaves of gold. It can be bought either as loose leaf or as transfer leaf. Typically, a book is about £40 depending on your supplier.

Above – gold leaf

The gold is approximately 10cm square.

The gilding process is straightforward, there are two types of gilding and these are transfer leaf and loose leaf.

Transfer leaf

This is the easier to apply of the two. The gold is adhered to the tissue paper in the book. A gold size (glue) is applied to the surface that you want to be gilded. This is a tinted varnish with a set drying time. The longer the drying time of the gold size the higher the shine you will get from the gold.

Gilding

The size is applied to the surface until it becomes tacky and then the gold is applied, and you rub the back of the tissue paper to transfer it to the surface. Once the size is dry, you can burnish the gold with some cotton wool so that you get a nice shine, and you remove any skewings or overlaps and loose gold.

Above - all the ornate work on the ceiling has been gilded

Loose leaf

This is a little trickier to apply. The gold is loose in the book and the slightest breeze will pick it up and take it floating away. For this reason, you need some specialist equipment.

Gilding

These are:

A gilder's cushion.

A pad with a paper "windbreak" to make the gold easier to handle. Some gilders do not use a cushion and will just gild straight out of the book.

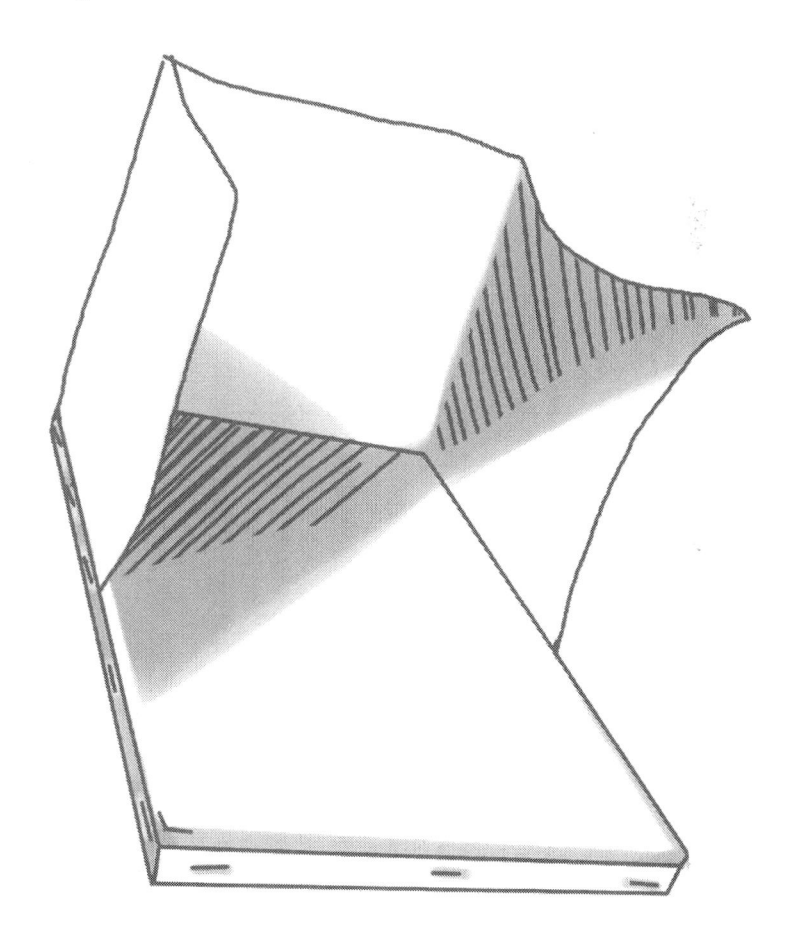

Above – a gilder's cushion

Gilding

A gilder's knife

This is a blunt but clean knife used to cut the loose gold into the size that you want. The knife must be kept clean otherwise it will not cut the gold cleanly.

Above – a gilder's knife

Gilding

A gilder's tip

This is a special brush designed to pick the gold up and transfer it to the surface being gilded. Usually glass.

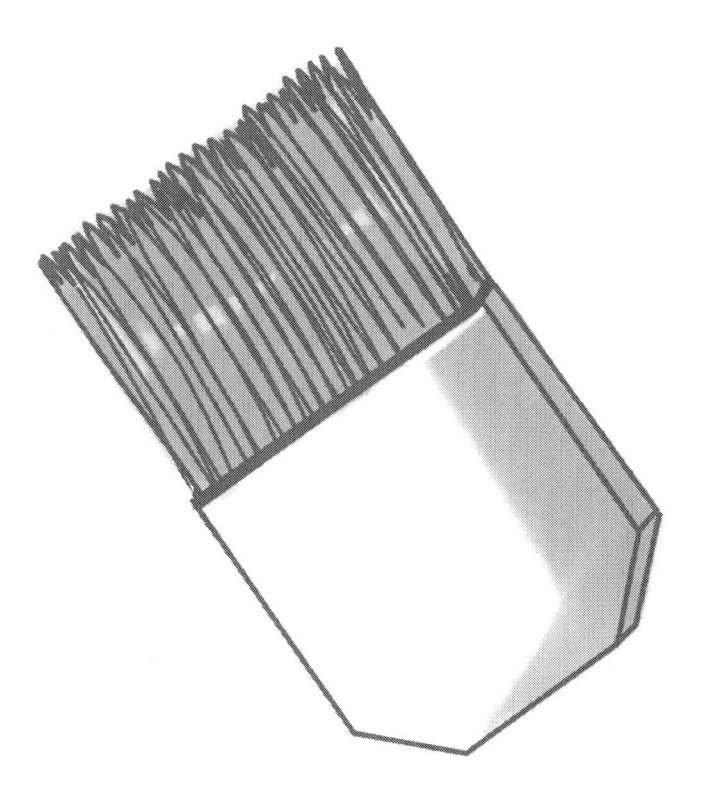

Above – a gilder's tip

Gilding

Loose leaf gold is used for gilding on glass, solicitors' windows and the like, the gold is applied to the inside of the window and viewed through the glass. The glue that is used therefore must be invisible because you are looking at the gold through the glass.

The gold size that is used for loose leaf is called gelatine size and comes in a capsule that is dissolved in water and flooded onto the glass using a brush called a gilders mop.

Above – a gilders mop

Gilding

Using the gilder's tip, you can pick up the gold, mainly by static. Once the gold gets near the glass it jumps from the tip to the glass and then smooths out as it hits the wet size on the glass.

Once dry, the gold is backed up with paint, usually gold paint and then black paint. This is to seal the gold onto the glass so that it can withstand the windows being cleaned. You could use gloss to back up the gold but, usually, signwriters enamel is used.

All gilding tools and gold itself can be bought from "Wrights of Lymm". Google them, all the goodies that we have spoken about are on there.

This is very specialist and it would take time to build up a portfolio of work. It can be very rewarding, and you would work on some very interesting projects. I did a fair amount of gilding in my twenties and worked on some really nice jobs.

You can charge a premium because the work is specialist, which is a bonus.

Gilding

Company name: The Gild Standard

Website: www.thegildstandard.co.uk

Tagline: Decorators who gild

Pitch: We have carried out specialist gilding work for churches, museums and several high-profile clients. We work to a very high standard and we work with gold. We thought about calling our company the gold standard but decided instead to focus on our gilding, not just with gold but with other precious metals too.

Section 6

Summing up

Chapter 28
The last one

The last one

Here we are at the end of the book. I hope you enjoyed our journey through the decorating specialisms that are out there.

When I initially had the idea for the book, I wrote down over 40 ideas but, as I got writing, I realised that if I wrote about all the ideas then the book would be like "War and Peace" and too thick!

I knew that you would not want this, so I have put some ideas on the back burner for now. I may do another book with more ideas another time. I will see what you think of this book first.

One of the main reasons that I have written this book is that I believe that we are lucky as decorators in that we have a great job and business. However, sometimes that does not seem to be the case.

A lot of guys wallow in misery and moan about low earnings and people that charge £80 a day and undercut them. There is nothing that you can do about what other people do but there is something you can do about what YOU do.

This is one of the main reasons for writing this book, to give you ideas for developing your business. You may want to focus completely on one specialism or you may keep doing what you are doing and develop a new specialism "on the side" so to speak, so that you develop additional revenue and security.

Here in the last chapter, I want to discuss three things on my mind at the moment and share them with you. These follow on from the specialism idea and hopefully will help you develop your business.

The last one

These are: the common pitfalls that we as decorators fall into, some top tips to build a bigger business and, finally, how we structure our time as self-employed business people.

The last one

Common pitfalls decorators fall into

I recently did a live webinar for "Trade Decorator" on pricing. Now, if you know me you will know that I am not a video person, I prefer to be locked away in a dark room writing about what I think.

However, if put under pressure I can turn on my "teacher mode" and fake it for an hour. This webinar was well received, and I had some great feedback from decorators.

One thing I discussed was the five common pitfalls that we as decorators fall into (myself included in the past) and I have included the points here so that they are forever available for you to look up when you feel you need to.

I feel it is important that we understand what the pitfalls are so that if we do decide to do them at least we are doing it with full awareness of the consequences. Some things I will have mentioned before in my other books, but this is a great summary and reminder for you.

1. Guessing the price

Oh yes, no surprise that this is my first pitfall. Many fall into this trap. We guess how long the job will take and then multiply this by our bargain basement day rate.

People are rubbish at guessing how long a task will take. This is well documented. Not just decorating but any task.

Try it yourself.

The last one

The next task that you do after reading this chapter: I want you to guess how long the task will take and then time yourself and see how far out you are. I bet it takes twice as long as you guess.

Let me give you an example. I am about to make myself a coffee. I think it will take me about 3 minutes. So, let's see how long it actually takes…

Okay, I am back, and the job actually took me 5 minutes. Oh, you say, you were only a couple of minutes out. But think about it, I took nearly twice as long as I guessed.

This was a short duration task as well; we are good at guessing common short duration tasks (like making a brew) longer tasks we are not so good at.

If you try to estimate a job that will take you a month then you will be miles off. If you overestimate and think it will take 6 weeks, then you will over price the job. If you underestimate the job and think it will take three weeks, then you will lose money.

The same goes for guessing materials.

If you currently guess then carry on for now but get a little black book, write all your guesses in there and then once you have done the task write the actual time taken. This information is valuable when working out your COST price.

More on that later.

2. Dropping the price (for no reason)

You have a look at a job that needs pricing. It's a new customer, let's call her "Mrs Jones", who has seen a post of your work on

The last one

Facebook and now wants your amazing skills at work on her lounge.

You measure up and discuss what she wants doing and then you go away and work out a price. Your price is £450. You telephone Mrs Jones to give her the price.

"Hello, Mrs Jones. It's Pete, the decorator. I came to look at your lounge last night," I say.

"Oh, hi Pete," she says.

"With all the materials included it will be £450," I say.

Silence…

"Mmmm," she says.

Silence again.

"It is a bit more than I was expecting," she says, "do you think you could do it for a little less?"

Right, let's stop for a second and have a think. You have three options here.

1. Drop the price to get the work.

2. Drop the price for a reason (e.g. prompt payment).

3. Don't drop the price and see if she still wants you to do the work.

Let us explore each option. If we just drop the price, she will think that the original price was an inflated one and she will also tell all her friends that you will drop your price if asked. Neither is something we want.

The last one

You could give her a 10% discount if she pays on the day of completion. That way you get something for the price drop. Alternatively, you could drop the price and maybe only put one coat on the walls and ceiling. Personally, I would not do either of those things.

I would expect to get paid promptly anyway and I would not want to risk the emulsion not covering and therefore making my work look poor to Mrs Jones' mates.

Finally, you could not drop the price. That way if you get the job it is worth doing and Mrs Jones gets to see that you are "business-like" and are not just making it up as you go along.

Back to the phone call.

"I have carefully calculated the price and £450 is what it will cost to get the work done. If you want a cheaper price maybe shop around and see what you can find. I will leave it with you, just let me know if you want to book me in. There would a 20% deposit of £90 when you book." I say.

That's it.

Leave it with her.

3. Working for anyone

When we first set up in business, we tend to work for our family and friends. Your gran, your mum and your mum's friends.

If you have set up in business for the long haul, then realise that one of your most important jobs is building up some great customers. These are the people that you want to work for and people that are willing to pay you what you are worth.

The last one

First, decide what you want your perfect customer is like. Here is mine:

Lives local
Will wait for my next available slot
Will pay my price
Will pay me a deposit
Is not a pain
Will pay me on completion

I carefully vet my new customers so that they meet these criteria. If you are not going to expand and you will stay as a one-man band then you will quickly get to the stage where you are not taking on any new clients.

What I have found is that the harder you are to engage, the more the customer wants you. So if you say that you are too busy to do their work and you don't even want to price it then you may find them begging for you to come and look at the work. From a negotiation point of view this is a great position.

4. Assuming the cheapest price is what everyone wants.

I tend to convert all situations into one that involves cars because I think it's a great way to understand the world. Plus, I like cars.

If you talk to most decorators, there is an assumption that all customers want to pay the cheapest price. If they get three quotes and you are one of them then you better be damn sure you are the cheapest.

Let us think about that for a second. I Googled what the cheapest car on the market today was (October 2020) and here

The last one

is what I found. The Dacia Sandero. Now don't be offended if you have one, it looks like a great car.

Above - The cheapest car available

But guess what? I don't know anyone who has one or wants one. Everyone I know has an Audi or a BMW or a Volkswagen. These are not the cheapest cars.

The last one

Above – what we really drive

People do not want the cheapest, they want the best value. That's a different thing. They fear cheap because they know that "buy cheap, buy twice". It is your job, therefore, to make sure that you are good value and that you educate the customer on what that is.

I read a post on Facebook the other day, it went a little bit like this.

A decorator (let's call him Dave) had priced a job for Mrs Smith and she had rung him up to say that she had a cheaper price from another decorator (let's call him Bob) and was there anything he could do with his price.

What does this tell you?

It tells you that Mrs Smith wants Dave to do the work at Bob's prices, otherwise why would she ring him up? I would just say to the customer that if you want the cheap price then get Bob to

The last one

do the work and if you want me to do the work then you will need to accept my price.

Let me know what you decide.

5. Doing the work for COST price instead of MARKET price

If you only remember one thing from this chapter, then this is the one.

The biggest pitfall that decorators make.

Before we start, how much is the car below? It is a Land Rover Evoke.

Above – The Land Rover Evoke

They are upwards of £31,000 so typically £35,000 to £40,000. This is the MARKET price for this car. We are all familiar with market price and what it means. I will come back to the Evoke later.

The last one

Everything we buy is priced at a level that the market will stand. Every business knows what it costs to make an item and also how much they can sell an item for.

My favourite example is keys.

A key will cost you £5 to get a new key cut and that is the market price. The cost price is about 25p. The key blanks are less than a penny if you buy them in bulk and the time to cut one is seconds.

COST price is time and materials = £0.25

MARKET price = £5.00

As you can see there is a big difference between the two.

What does a decorator do when he prices a job? He works out his time, let's say £120 per day (this just about covers wages) and he works out his materials.

Like this:

5-day job @ £120 per day = £600

Materials = £120

Overheads = £20

Total COST = £740

This is the **cost** price. It is the same as the 25p that the key cutter calculates. Does the key cutter sell his keys at 25p? No, he doesn't, and we should not sell our service for cost either.

We need to do two things.

The last one

We need to know what the market price for our work is. This is the average that decorators that do your type of work charge in your area.

You need to get your cost price down as low as you can by being more efficient.

I will constantly ask decorators and decorating companies what they a charge for a range of work. That way, I have a good idea of what the market price is. I like to break things down to a metreage rate so that we can have a meaningful conversation with other decorators.

For example:

If I want £120 per day and I paint a 6m by 4m ceiling, that's 24m^2 and I prepare and paint it, two coats in half a day then my meterage rate is

£60 (half a day) / 24 (area of ceiling) = £2.50 per metre

Materials would be about £30

£30 / 24 = £1.25

To summarise:

Time = £2.50

Materials = £1.25

Overheads = 50p

(My overheads are £2.50 per hour, 4 hours to do the ceiling is £10 so £10 divided by 24m^2 is 50p per metre)

Total = £4.25 COST PRICE per metre

The last one

This is our cost price per metre, it's ours too, no-one else's. If I do a ceiling in 2 hours and not 4, then my cost price is even lower.

This now enables me to discuss price with everyone. If a builder says that he will pay me £4.50 a metre to emulsion a ceiling, then I know that I can make money at this. But not much (25p per metre).

Many people I speak to charge between £4 and £8 per metre for emulsion so now that means something to me. I can also work on getting the cost price down by doing it faster. So, if I prep and spray a ceiling (two coats) in an hour then my cost price is now just over a pound per metre.

Suddenly, I am making good money, especially if I charge £6 per metre.

Let us get back to the Evoke. How much do you think it costs to make an Evoke in the factory? I will tell you: £3,000.

If decorators made Evokes, they would probably sell them for £3,000 instead of £40,000.

No wonder we never have any money!

The last one

Building a bigger business

Before I start with this topic, I know that not everyone wants to build a big business. Many are happy on their own or with just three or four in a team.

You may have tried to go bigger and found it a headache, so you have scaled back.

The reason that decorators struggle with going bigger is that they don't do it correctly. Now building a bigger business is a book in itself, so I will not go into it in great detail here.

I will give you nine things that you need to do if you want to build a bigger business without the headache.

1. It must be profitable, ideally COST plus 50%

To build a business you need extra money over and above your wages, overheads and your materials. You need money to invest back into the business and grow. You need a marketing budget; you need money to spend on time off the tools planning and meeting new customers.

I think we feel guilty if we make too much money, but we shouldn't. If you build a great company then you will be employing lots of people and helping them to pay their mortgages and feed their families.

2. You must have a SIMPLE and RELATEABLE differentiator

The whole book has been about this. Making your business different from the rest. But keep it simple so that the customer instantly understands what makes you different. Remember "Just Dents" – simple and relatable.

People need to relate to what you do and instant "get" the problem you are solving without a big explanation. "Our finish is like a car respray" for example. Easy to relate to.

3. You need to operate in a large enough market

This is a harder one to understand. If you specialise then the tighter the specialism, the bigger the area that you will need to serve to keep growing.

Some internet companies have a global market so they can specialise narrow. You could have the whole of the UK as a market, which in fairness is big. You may have a broader specialism and a smaller area. The North West of England, for example.

4. Focus on consistent revenue with high lifetime customer value

Sometimes this is where we go wrong as decorators. We look at the price of a job and we look at a customer as a one-off. This is because typically we may only visit a repeat customer once a year. If you are selling a tin of baked beans, then a customer will buy your product every week, so the lifetime value becomes more obvious.

Let's look at an example. You get a phone call from a customer and they want you to look at painting their front door. They are a new customer, and you are busy. Your first thought would be "is it worth going to look at the job just for a front door?"

Before we answer this question, let's look at it from the customer's point of view. The customer is the managing director of a distribution company. They have a few warehouses and

The last one

property. Their house is large, and they need the whole lot decorating.

Their old decorator who they liked has retired and they are looking for a new one. The front door job is a test.

What will you price it? What are you like? Will you do a good job even though it's only a small job? Will you be prompt, clean and tidy?

This customer could easily spend £2,000 a year with you. If you work for them for 40 years, then that's £80,000. So, the front door could be worth a lot of money in the future.

As a business, you are looking to build customers who will keep coming back and spending money with you. That way your business will grow based on a nice solid customer base.

5. Invest in a memorable brand

We have explored this in Section 1, so there is no need to go over it again except to say that the brand is what has value. If you build a company known for doing a certain kind of work and that name is known widely then this has a lot of value. Do not underestimate the power of this for marketing purposes.

6. Start a team

When you set up as a decorator, you need to make a decision. Will you stay small (on your own) or will you build a business? If you want to build a business, then you will need a team.

We all have our strengths and weaknesses and for your company to grow and succeed you will need people that can do all the things needed.

The last one

You need an office person to send out quotes and invoices, chase money and keep records up to date. These days you do not have to employ an office person full time or even have an office. You can hire a virtual assistant (VA) on an hourly rate and just pay them for what they do.

It would be a good idea to hire a marketing person, too. All the best businesses have people that complement each other. The classic example is Apple with Steve Wozniak and Steve Jobs. Jobs was the business and marketing brains, and Woz was the technical talent.

7. Know how to say "No"

This is hard in the early days because you want to please everyone, but if you want to develop you need to turn away work that is not a good match for what you are trying to do.

8. You need monthly growth

This is obvious really but if you are a small business now and you want to be a big business then you need to grow. How fast you grow is up to you, but you are better growing consistently.

Let us look at an example.

You have just started and you do £2,000 worth of work in your first month. This is turnover and not profit so this figure is conservative. I wanted to start low.

We decide that we will grow by 10% a month, this means that in the second month we only need to do £200 (10% of £2,000) extra work, making the turnover in month two £2,200.

The last one

Month three would need an extra £220 worth of work (10% of £2,200) and this would be £2,400. Still low.

If you continued to grow at this rate, at the end of the first year you would be turning over £68,000. Average for your first year, I would think.

By the end of the second year, you would be turning over £200,000. This is more serious but all you have done is consistently grown by just 10% a month.

By the end of the third year, your turnover would be £700,000. By now, you have a decent team, your marketing process is in full swing and you employ someone full time to do this.

By the end of year four, your turnover is £2.1 million. Yes, I know it seems to scale up quicker and quicker, but this is the beauty of having a consistent growth target. Perhaps 10% a month is too much, and you just want to grow by 5% a month. That is fine, it's slower but you will still get there in the end.

9. Think long-term

This is an important one. Sometimes we do things from week to week and forget the long-term. A lot of things are good for the long-term but have no immediate benefit. I can give you a good example from my own business.

Back in 2016, I did some onsite training in spraying for local people. To help with this I wrote a book on airless spraying to give out after the training.

What are the immediate benefits of this? Well, there are none. It took a lot of time to sit down and do it. I did not get paid for this time. I wrote in the evening so, when I got home from work,

The last one

the last thing that I felt like doing was writing about spraying. This went on for months. I was doing something that had no short-term benefit.

Fast forward to the present day and, because of that long-term investment in my business, I get many offers of interesting work. Too much really. So now I am getting the rewards of that long-term thinking.

This applies to all aspects of your business but is things like training (yourself and your team), marketing, testing products, networking, developing new processes and building relationships with customers.

The last one

Structuring our time

When we work for a company, they do all the thinking for us. They tell us when to start in the morning, how long we can have for our lunch and what time we can go home. They tell us how many days of holiday we can have and how many days a week we can work.

Typically, if you work for a company you will work for 20 days a month and have 8 days off (the weekends). You would also work a 40-hour week.

One thing I have struggled with myself being self-employed is organising my time. Somehow, being able to do whatever you want makes it harder and not easier. I have spoken to a few decorators about this and we fall into 1 of 4 camps.

1. Stick to the old employee model

I think this is most of us. We carry on with the old model. We work from 8 am until 5 pm and try to have weekends off so we can recover. Seems a shame really, we may as well just stay employed.

2. Work every hour on production

Probably even more likely, you work 7 days a week decorating trying to keep all those customers happy. We never get time for a break and we never get time to spend on developing ourselves or the business and because of this our business goes nowhere.

The last one

3. Work every hour on business development and production

There comes a time when you decide that if you want to go somewhere with your decorating business you need to do some development.

First, you need to develop yourself. The skills needed to be a decorator differ from the skills needed to run a team of ten decorators. You need to learn those skills.

Second, you need to develop your business. These are things like expanding your customer base, recruiting new decorators, and developing your systems of work. All these things sometimes get put on the back burner because they are not urgent, and you just bumble along getting nowhere.

So, you may work 5 days a week decorating, 2 days a week developing your business and every evening pricing jobs.

This is still no good. You need time to recharge your batteries. I have thought about this a lot and the guys that ran the factories in the past knew how to make money and they would not have given you 2 days off a week if it didn't benefit them.

They knew that you needed a rest and that you needed interests outside work so that when you were at work you produced loads of stuff.

We tend to overlook the importance of rest and recuperation. I know I do. I plan a few days "off" and a dream job comes along that I can't say "No" to and away I go. But I am wrong to do this. The days off are more important than we realise.

The last one

4. A nice balance of production, business development and time to recharge

This is what I am trying to do at the moment. I call it the "Ten Triangle".

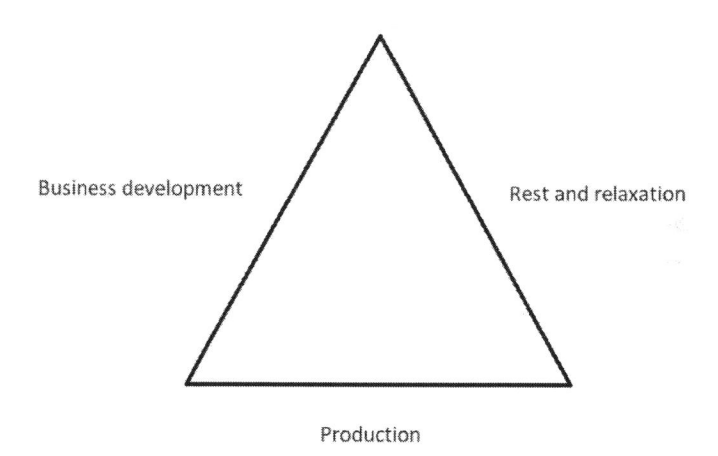

```
                    /\
                   /  \
                  /    \
                 /      \
Business development   Rest and relaxation
               /          \
              /            \
             /_____\
                Production
```

Now, you must excuse me for the made-up term "Ten triangle" and the diagram, but I think I have read too many management books and I feel I need to have my own theory to put out there. In fairness, I think it explains the idea well.

This is a three-way split of your time and you may not get to the ideal split straight away, so it is just a work in progress.

A third of your time working to earn cash, a third of your time developing your business and finally a third of your time resting and living your life.

The last one

If we look at a typical month of 30 days that means 10 days decorating, 10 days developing your business and 10 days resting.

10 days a month resting seems like a luxury but if you consider that you had 8 days when you worked for your old firm (four weekends) then it is not too much to ask for an extra 2 days a month now that you have your own business.

10 days a month developing your business seems too much, but again it's surprising how much development tasks take. For example, you may do some training for 2 days a month. You may have 2 days "selling" and setting up meetings with new customers. You may have 2 days meeting customers and pricing.

There are many development tasks that you could do but don't because you do not have the time. For example, reading this book is development. Spending time building a network on LinkedIn is development.

Ten days a month decorating does not seem enough and, in fairness, it depends how much you earn per day and how much you need. If you can draw £200 per day, then this brings in a healthy £2,000 a month. I know decorators that draw less to minimise their tax liability and leave as much in the business as they can. This is something that you need to decide.

You may not follow the 10-day x 3-way split, but still follow the triangle model. You may do 20 days decorating, 5 days developing and 5 days off. You need to be working towards more time recharging your batteries and spending time with your family, more time developing your business too, so that eventually you will be free to work when you want.

The last one

I have been trying to do this myself and I have to admit it's hard to stick to it. I tend to skip the days off and work. I am good at development and I am doing more and more of this these days.

I do roughly 14 days production, 6 days development and 8 days off.

I will finish here and have a rest for a couple of days, even though it's Wednesday today. These days I wake up and I am not sure what day it is!

I hope you have enjoyed the book; I do think that being a specialist decorator is the future and that the ones that embrace this approach will be the decorators that prosper. You only need to look at trends in other countries such as America and Australia. The younger decorators that I speak to seem to get it and I think we are moving into a golden age for decorators.

I hope so.

Other books by the author

Fast and Flawless

A guide to airless spraying

This is a chatty guide to airless spraying for decorators, decorating students and anyone interested in spraying with an airless system. The book covers all aspects of the airless sprayer including the components of the system, the different systems that are out there to buy and setting up the system.

The book covers topics such as types of sprayers, essential equipment, using the equipment, masking, PPE and masks, a bit about paint, what to do when it all goes wrong, spraying in the real world and common paint defects.

Other books by the author

Fast and Flawless Pricing

A guide to pricing and business for decorators

Are you a decorator that struggles with pricing?

Have you just set up in business and are looking for some pointers?

Are you an established business looking for some inspiration on how to move forward?

This chatty guide on pricing and business will gently guide you through the process of pricing a decorating job. It looks at the pitfalls of getting your pricing wrong and the advantages of having a good pricing system.

The book has been written by someone who has both been a decorator and taught decorating in a local college for most of his life.

Other books by the author

Fast and Flawless Systems – A decorator's guide to planning and carrying out a successful job

This book looks at systems for decorators.

This book covers all types of systems from which paint to use on what surface to what order you should spray a room. The book also covers aspects of decorating that you may or may not be aware of such as painting uPVC, training, funding and marketing.

If you have read the other two books already then this is one is a must-read. If you haven't, then this book is a great place to start.

So, do you want to be a decorator?

An insight into the varied careers for a decorator.

The author has had many years of experience in the decorating industry and has written several books on subjects such as spraying, pricing, and planning your jobs.

If you are a decorator who is just starting out in your career, then you may be wondering what the options are for you in the future if you want to develop your career.

You may be a decorator who has been doing the job for many years and you are thinking that maybe you would like to earn more money or simply have a change from where you are currently working.

Whatever stage you are at with your career, this book will act as a guide to show you what is out there and what steps you need to take to make the change to a new job.

The book discusses a range of career choices that are available to the decorator. Maybe you are an apprentice working for a one-man band and you fancy working for a larger decorating company. Maybe you have been decorating for a while and you fancy doing some teaching. Maybe you want to increase your earnings and become a site manager.

All these options and more are explored.

Other books by the author

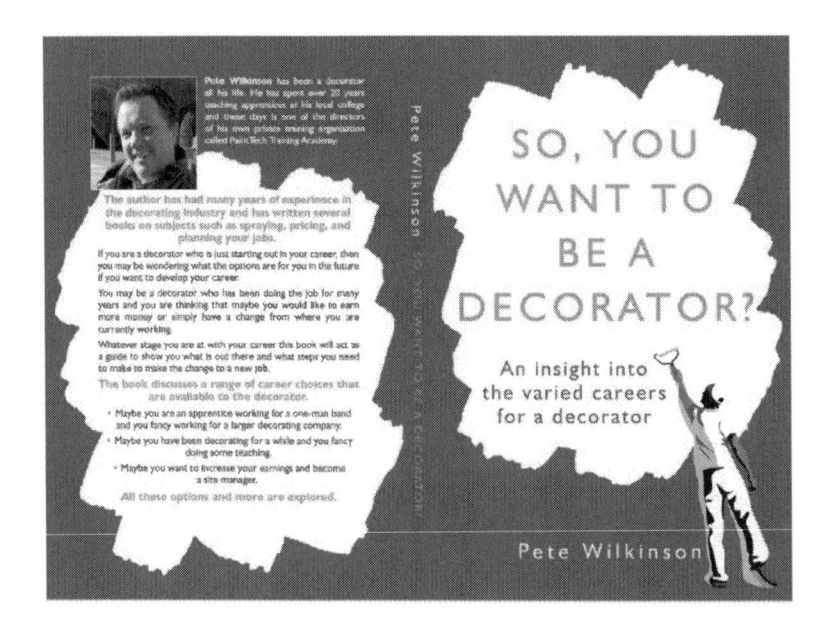

SO, YOU WANT TO BE A DECORATOR?

An insight into the varied careers for a decorator

Pete Wilkinson

Pete Wilkinson has been a decorator all his life. He has spent over 20 years teaching apprentices at his local college and these days is one of the directors of his own private training organisation called PaintTech Training Academy.

The author has had many years of experience in the decorating industry and has written several books on subjects such as spraying, pricing, and planning your jobs.

If you are a decorator who is just starting out in your career, then you may be wondering what the options are for you in the future if you want to develop your career.

You may be a decorator who has been doing the job for many years and you are thinking that maybe you would like to earn more money or simply have a change from where you are currently working.

Whatever stage you are at with your career this book will act as a guide to show you what is out there and what steps you need to make to make the change to a new job.

The book discusses a range of career choices that are available to the decorator.

- Maybe you are an apprentice working for a one-man band and you fancy working for a larger decorating company.
- Maybe you have been decorating for a while and you fancy doing some teaching.
- Maybe you want to increase your earnings and become a site manager.

All these options and more are explored.

Other books by the author

Tales from the building site

Lessons learned when working on a big site

The author has spent many years working in the building trade as a decorator. During all those years he has seen things that have made him laugh and things that have made him tear his hair out.

There have also been many occasions that have made him proud to be part of it all. Here is a book for all you people in the trade and also for everyone else who wonders what goes on behind those big high hoardings that clearly state the public are not allowed in.

It is a warts and all look behind the curtain from the perspective of a decorator. Be prepared to be shocked, to laugh and to shake your head in disbelief.

Other books by the author

Boat Life

The trials and tribulations of living aboard

Nothing to do with construction or decorating. I love boats, I have one and I have lived aboard myself, so this is an insight into the lifestyle.

This is a book for Boaters, written by a Boater. Pete Wilkinson has spent his whole life around boats and has owned a couple too.

The book looks at all aspects of boating including, what is the best boat to buy, where to look when buying a boat and do you build one or do you buy one?

The following questions are answered: Which is the best type of boat – wood, fibreglass or steel? Do you borrow

Other books by the author

money or save to get your boat? Do you move around or stay put? What is the essential kit needed? How do you keep her shipshape?

The book also looks at living aboard and gives an insight into the liveaboard life. The book also puts boating into the context of modern life and discusses the advantages of living onboard.

Finally, if you have wondered what goes on behind the curtain of a Boater's life then this book will show you.

Other books by the author

Check out the website

If you are interested in being kept up to date with future books, or you just fancy the odd freebie, then subscribe on my website.

www.fastandflawless.co.uk

About the author

Pete Wilkinson has been a decorator all of his life. In his younger years, he worked for a medium-sized decorating company doing a wide range of work.

Then at the age of 27, he got a job teaching Painting and Decorating at a local college. These days he runs his own training company called PaintTech Training Academy.

When he is not working, he likes to spend time relaxing on his boat with his wife Tracey.

Printed in Great Britain
by Amazon